# Cellular Responses to Arsenic: DNA Damage and Defense Mechanisms

*The mission of the Awwa Research Foundation (AwwaRF) is to advance the science of water to improve the quality of life. Funded primarily through annual subscription payments from over 1,000 utilities, consulting firms, and manufacturers in North America and abroad, AwwaRF sponsors research on all aspects of drinking water, including supply and resources, treatment, monitoring and analysis, distribution, management, and health effects.*

*From its headquarters in Denver, Colorado, the AwwaRF staff directs and supports the efforts of over 700 volunteers, who are the heart of the research program. These volunteers, serving on various boards and committees, use their expertise to select and monitor research studies to benefit the entire drinking water community.*

*Research findings are disseminated through a number of technology transfer activities, including research reports, conferences, videotape summaries, and periodicals.*

# Cellular Responses to Arsenic: DNA Damage and Defense Mechanisms

Prepared by:
**X. Chris Le**
Department of Public Health Sciences, Faculty of Medicine, University of Alberta
Edmonton, Alberta, Canada T6G 2G3

and

**Michael Weinfeld**
Experimental Oncology, Cross Cancer Institute, Edmonton, Alberta, Canada T6G 1Z2

Jointly sponsored by:
**Awwa Research Foundation**
6666 West Quincy Avenue, Denver, CO 80235-3098

and

**U.S. Environmental Protection Agency**
Washington D.C.

and

**Association of California Water Agencies**
Sacramento, CA

Published by:

Awwa Research Foundation
Advancing the Science of Water®

IWA Publishing

## DISCLAIMER

This study was funded by the Awwa Research Foundation (AwwaRF). AwwaRF assumes no responsibility for the content of the research study reported in this publication or for the opinions or statements of fact expressed in the report. The mention of trade names for commercial products does not represent or imply the approval or endorsement of AwwaRF. This report is presented solely for informational purposes.

**Published by IWA Publishing, Alliance House, 12 Caxton Street, London SW1H 0QS, UK**

Telephone: +44 (0) 20 7654 5500; Fax: +44 (0) 20 7654 5555; Email: publications@iwap.co.uk;
Web: **www.iwapublishing.com**

AwwaRF report number 90976F. Originally published by AwwaRF for its subscribers in 2003.
IWA Publishing version published 2004.



ISBN 1 84339 880X

## CONTENTS

LIST OF TABLES ........ ix
LIST OF FIGURES ........ xi
FOREWORD ........ xiii
ACKNOWLEDGMENTS ........ xv
EXECUTIVE SUMMARY ........ xvii
CHAPTER 1: INTRODUCTION ........ 1
Arsenic ........ 1
Cellular Defense Mechanisms ........ 3
Assays for DNA Damage ........ 7
CHAPTER 2: GENETIC PREDISPOSITION TO THE CYTOTOXICITY OF ARSENIC: THE ROLE OF ATM AND XRCC1 ........ 9
Introduction ........ 9
Materials and Methods ........ 10
Cell Lines and Culture Conditions ........ 10
Treatment with Sodium Arsenite ........ 11
Cell Irradiation ........ 11
Colony-Forming Assay ........ 11
Measurement of Thymine Glycol ........ 11
Measurement of DNA Double-Strand Breaks (Phosphorylated Histone H2AX) ........ 12
Western Blotting Assay ........ 12
Cell Cycle Analysis ........ 13
Results ........ 13
Induction of Thymine Glycol (Tg) by Sodium Arsenite ........ 13
Dose and Time Dependence of Sodium Arsenite-Induced Cytotoxicity of A549 Human Cells ........ 14
Comparison of the Cytotoxic Effects of Four Arsenic Compounds ........ 14
Cytotoxic Response of DNA Repair Deficient Cell Lines Towards Sodium Arsenite ........ 15
Response of AT Cells to Sodium Arsenite ........ 16
Response of Double-Strand Break Repair Deficient Cells to Sodium Arsenite ........ 17
DNA Double-Strand Break Induction by Sodium Arsenite ........ 19
Changes in p53 Protein Levels in Response to Sodium Arsenite ........ 20
Influence of Sodium Arsenite on the Cell Cycle ........ 21
Discussion ........ 21
CHAPTER 3: ADDITIVE EFFECTS OF SODIUM ARSENITE AND γ-RADIATION ........ 25
Introduction ........ 25
Results ........ 25
Effect of $As^{III}$ on the Removal of Thymine Glycol in A549 Cells ........ 25
Cytotoxicity of Human Cell Lines to $As^{III}$ and γ-Radiation as Single Agents ........ 26
Cytotoxicity of the Combination of $As^{III}$ and γ-Radiation ........ 27

Comparison Between Pre- and Post-Treatment with $As^{III}$ in γ-Irradiated Cells ........ 31
Effects of Low Concentrations of $As^{III}$ ........ 32
Discussion ........ 33
CHAPTER 4: ASSAY FOR DNA DAMAGE USING CAPILLARY ELECTROPHORESIS LASER-INDUCED FLUORESCENCE ........ 35
Introduction ........ 35
Materials and Methods ........ 37
Reagents ........ 37
Design of Probe ........ 37
Synthesis of Damaged Oligonucleotide ........ 39
Purification of BPDE-Oligonucleotide ........ 39
Synthesis and Purification of 90-mer Oligonucleotides ........ 40
Instrumentation for Analysis of Ligation Products ........ 40
Characterization of BPDE and Control 90-mers ........ 41
Treatment of A549 Cells with BPDE ........ 43
Competitive Assay for BPDE-DNA Adducts ........ 43
Results and Discussion ........ 43
Purification of BPDE-16-mer Oligonucleotide ........ 43
Synthesis and Purification of BPDE 90-mer Ligation Products ........ 47
Characterization of the BPDE 90-mer Ligation Products ........ 47
Affinity Interaction of BPDE-90-mers with a Monoclonal Antibody ........ 50
Determination of Specific Antibody Using BPDE-90 mer as a Probe ........ 53
Screening for anti-BPDE Antibodies Using the Fluorescent BPDE-90-mer Probe ........ 55
Application of the BPDE-DNA Probe to Competitive Assay for BPDE-DNA Adducts in Cells ........ 58
CHAPTER 5: EFFECTS OF ARSENITE AND BPDE IN THE INDUCTION AND REPAIR OF BPDE-DNA ADDUCTS IN A549 CELLS ........ 61
Introduction ........ 61
Materials and Methods ........ 61
Reagents ........ 61
Preparation of 190-mer Oligonucleotide Standard ........ 62
Instrumentation ........ 63
Treatment of A549 Cells with BPDE and Arsenite ........ 63
Sample Preparation ........ 64
Analysis with CE-LIF ........ 64
Results and Discussion ........ 65
Capillary Electrophoresis Immunoassay for BPDE-DNA Adducts ........ 65
Assay for BPDE-DNA Adducts in Cellular DNA ........ 68
Dose Responses of BPDE and BaP ........ 70
Removal of BPDE Adducts from the DNA of A549 Cells Exposed to 1 μM BPDE ........ 73
Removal of BPDE Adducts from the DNA of A549 Cells Exposed to 1 μM BPDE and 100 μg/L $As^{III}$ ........ 74

Comparison of A549 Cells Exposed to 1 μM BPDE Alone and In Comparison with 100 100 μg/L $As^{III}$ .......... 75
CHAPTER 6: ARSENIC SPECIATION ANALYSIS AND SPECIES STABILITY .......... 79
Introduction .......... 79
Instrumentation .......... 80
HPLC Separation with Hydride Generation Atomic Fluorescence Detection (HPLC-HGAFS) .......... 80
HPLC-Inductively Coupled Plasma-Mass Spectrometry (HPLC-ICPMS) .......... 81
Standards, Reagents, and Samples .......... 82
Procedures .......... 83
Speciation of $As^{III}$, $As^{V}$, $MMA^{V}$, and $DMA^{V}$ in Urine .......... 83
Speciation of $As^{III}$, $As^{V}$, $MMA^{V}$, $DMA^{V}$, $MMA^{III}$, and $DMA^{III}$ in Urine .......... 83
Determination of TMAO .......... 83
Sample Storage Experiments .......... 83
Results and Discussion .......... 84
Speciation of the Usual Arsenic Compounds in Human Urine .......... 84
Speciation of Intermediate Arsenic Metabolites in Human Urine .......... 87
Determination of TMAO .......... 90
Stability of $MMA^{III}$ and $DMA^{III}$ .......... 92
Stability of $MMA^{III}$ in Deionized Water .......... 92
Stability of $MMA^{III}$ in Urine .......... 93
Stability of $DMA^{III}$ in Deionized Water .......... 98
Stabillity of $DMA^{III}$ in Urine .......... 98
Preservation of $MMA^{III}$ and $DMA^{III}$ .......... 98
CHAPTER 7: INTERACTION OF ARSENICALS WITH METALLOTHIONEIN .......... 101
Introduction .......... 101
Materials and Methods .......... 101
Reagents .......... 101
Instrumentation .......... 102
HPLC-ICPMS .......... 102
Triple Quadrupole Time-of-Flight Mass Spectrometry .......... 102
Procedures .......... 102
Results and Discussion .......... 103
Binding of Arsenic to Metallothionein .......... 103
Binding of $MMA^{III}$ and $DMA^{III}$ to Metallothionein .......... 107
Discussion and Concluding Remarks .......... 114
REFERENCES .......... 117
ABBREVIATIONS .......... 145

## TABLES


| | | |
|---|---|---|
| 2.1 | Fluorescence intensity of phosphorylated histone H2AX | 19 |
| 3.1 | Parameters determined for the interaction of 2-Gy irradiation with $As^{III}$ exposure | 30 |
| 4.1 | Comparison of denatured and native BPDE-90-mers on their affinity with anti-BPDE antibody 8E11 | 53 |
| 5.1 | Summary of measured BPDE-DNA adduct levels from various experiments | 73 |
| 5.2 | Comparison of peak areas of antibody-DNA adduct complex from the analysis of cellular DNA from A549 cells either treated with BPDE only or with a combination of BPDE and $As^{III}$ | 76 |
| 6.1 | ICPMS Operating Conditions | 82 |
| 6.2 | Residual amounts of $MMA^{III}$ (μg/L) remained in the samples after each sample was spiked with 100 μg/L $MMA^{III}$ and the samples stored for up to 4 months | 92 |
| 7.1 | Theoretical and measured molar masses of the MT and MT-$As^{III}$ species | 106 |
| 7.2 | Expected and experimental masses of arsenic-related amino acid species observed in tandem mass spectra | 107 |
| 7.3 | Expected and experimental molecular masses for unbound MT and MT bound with $MMA^{III}$ | 110 |
| 7.4 | Expected and experimental molecular masses for unbound MT and MT bound with $DMA^{III}$ | 110 |

## FIGURES


1.1 Biomethylation of arsenic showing a sequence of reduction of pentavalent to trivalent arsenic followed by oxidative addition of a methyl group .................................... 2
1.2 Schematic representation of base excision repair (BER) ........................................ 4
1.3 Model for nucleotide excision repair (NER) in mammalian cells showing damage recognition, opening of a region around lesion, dual incision, repair synthesis ....... 6
2.1 Oxidation of thymine to thymine glycol ............................................................... 9
2.2 Generation of thymine glycol in the DNA of human A549 cells by cellular exposure to sodium arsenite for 24 h ..................................................................... 13
2.3 Cytotoxicity of sodium arsenite towards A549 human lung carcinoma cell line ..... 14
2.4 Comparison of the cytotoxic effects of 24 h exposure to increasing doses of $As^{III}$, $As^{V}$, $MMA^{V}$, and $DMA^{V}$ with A549, GM43, and CRL-1223 cells ............ 15
2.5 Response of Chinese hamster ovary cells to 24 h exposure to sodium arsenite ....... 17
2.6 Response of double-strand break repair deficient cells to 24 h exposure to sodium arsenite and $\gamma$-radiation ........................................................................ 18
2.7A Western blot analysis of p53 protein levels in AT and repair proficient cells after exposure to sodium arsenite ........................................................................ 20
2.7B Western blot analysis of phosphorylation of p53 serine 15 in A549 cells in response to exposure to $As^{III}$, $\gamma$-radiation, and UV radiation .............................. 20
2.8 The effect of $As^{III}$ on cell cycle distribution of GM43 and AT5BI cells ................. 21
3.1 Effects of $As^{III}$ on the removal of thymine glycol in A549 cells ............................ 26
3.2 Cell survival of various cell strains after treatments ............................................. 27
3.3 Effects of $As^{III}$ on 2-Gy $\gamma$-irradiated cells of different cell lines ............................ 28
3.4 Effect of 300 $\mu$g/L $As^{III}$ on $\gamma$-irradiated cells......................................................... 29
3.5 Effects of 300 $\mu$g/L $As^{III}$ on $\gamma$-irradiated GM38 and CRL-1223 (XP-A) cells ......... 31
3.6 Comparison of cell survival of pre- vs. post-treatment with $As^{III}$ for 1 day ............ 32
3.7 Effects of different concentrations and times of $As^{III}$ on $\gamma$-irradiated CRL-1223 (XP-A) cells .................................................................................................. 33
4.1 Schematic representation of the design of the DNA damage probe ........................ 38
4.2 Schematic showing capillary electrophoresis with laser-induced fluorescence detection system.................................................................................................... 42
4.3 First step in the HPLC purification of BPDE-modified 16-mers from a reaction involving *anti*-BPDE and a 16-mer oligonucleotide .............................................. 45
4.4 Second step in the HPLC purification of BPDE-modified 16-mers ........................ 46
4.5 Effect of heat denaturation on BPDE-modified and control 90-mers ...................... 49
4.6 Capillary electrophoresis analysis of the fluorescent 90-mers and their mixtures with antibody 8E11 ............................................................................................. 51
4.7 Effect of overnight incubation for BPDE 90-mer and 8E11 antibody ..................... 52
4.8 Incubation of BPDE 90-mer with varying concentrations of antibody 8E11 .......... 54
4.9 Comparison between BPDE-DNA antibodies using the BPDE 90-mer fluorescent probe ................................................................................................... 57
4.10 Representative electropherograms showing competitive assay for BPDE-DNA adducts from A549 cells ...................................................................................... 59

5.1 Schematic diagram showing the preparation of a 190-mer containing a single BPDE adduct ........ 63
5.2a Illustration of immuno-reaction in the capillary electrophoresis immunoassay ....... 65
5.2b Illustration of CE/LIF analysis of immuno-complex with DNA adduct in capillary electrophoresis immunoassay ........ 66
5.3a Series of electropherograms from the CE/LIF analysis of the mixture of various concentrations of BPDE-190-mer incubated with 2 and 1 antibodies ........ 67
5.3b,c Relative peak area and peak height of the complex of BPDE-190-mer with 2 and 1 antibodies as a function of concentration of BPDE-190-mer ........ 68
5.4 Schematic illustration of analysis of adducts in cellular DNA ........ 69
5.5 Series of electropherograms from CE/LIF analysis of BPDE-DNA adducts from A549 cells incubated with B(a)P ........ 70
5.6 Dose-response curves from A549 cells incubated with BPDE and B(a)P ........ 72
5.7 Peak areas of antibody-DNA adduct complex from analysis of cellular DNA ........ 74
5.8 Peak areas of antibody-DNA adduct complex from analysis of cellular DNA ........ 75
6.1 Schematic diagram showing the HPLC-HGAFS system for arsenic speciation analysis ........ 80
6.2 HPLC-HGAFS analyses of As$^{III}$, As$^{V}$, MMA$^{V}$, and DMA$^{V}$ ........ 86
6.3 Typical chromatograms showing HPLC-HGAFS analyses of As$^{III}$, As$^{V}$, MMA$^{V}$, DMA$^{V}$, MMA$^{III}$, and DMA$^{III}$ ........ 88
6.4 Chromatorgrams showing speciation analyses of arsenic ........ 89
6.5 Chromatograms showing HPLC-HGAFS analyses of TMAO ........ 91
6.6 Effect of storage duration and temperature on the stability of MMA$^{III}$ in deionized water ........ 94
6.7 Typical chromatograms from the HPLC-HGAFS analysis of deionized water samples ........ 95
6.8 Effect of storage duration and temperature on the stability of MMA$^{III}$ in urine ........ 96
6.9 Chromatograms from the HPLC-HGAFS analysis of urine samples spiked with MMA$^{III}$ ........ 97
6.10 Effect of storage duration and temperature on the stability of DMA$^{III}$ in deionized water ........ 99
6.11 Effect of storage duration and temperature on the stability of DMA$^{III}$ in urine ........ 100
7.1 HPLC/ICPMS analysis of MT-As$^{III}$ and MT-MMA$^{III}$ complexes ........ 104
7.2 ESI mass spectra from the analysis of solutions containing 7 $\mu$M MT and varying amounts of As$^{III}$ ........ 105
7.3 Schematic representation of the binding stoichiometry between MT and As$^{III}$, MMA$^{III}$, and DMA$^{III}$ ........ 109
7.4 ESI mass spectra from the analysis of solutions containing 7 $\mu$M MT and varying amounts of MMA$^{III}$ ........ 111
7.5 ESI mass spectra from the analysis of solutions containing 7 $\mu$M MT and varying amounts of DMA$^{III}$ ........ 112
7.6 Partial ESI/MS/MS spectra showing the low mass region for MT[As(CH$_3$)]$_7$ and MT[As(CH$_3$)$_2$]$_7$ ........ 113

## FOREWORD

The Awwa Research Foundation is a nonprofit corporation that is dedicated to the implementation of a research effort to help utilities respond to regulatory requirements and traditional high-priority concerns of the industry. The research agenda is developed through a process of consultation with subscribers and drinking water professionals. Under the umbrella of a Strategic Research Plan, the Research Advisory Council prioritizes the suggested projects based upon current and future needs, applicability, and past work; the recommendations are forwarded to the Board of Trustees for final selection. The foundation also sponsors research projects through the unsolicited proposal process; the Collaborative Research, Research Applications, and Tailored Collaboration programs; and various joint research efforts with organizations such as the U.S. Environmental Protection Agency, the U.S. Bureau of Reclamation, and the Association of California Water Agencies.

This publication is a result of one of these sponsored studies, and it is hoped that its findings will be applied in communities throughout the world. The following report serves not only as a means of communicating the results of the water industry's centralized research program but also as a tool to enlist the further support of the nonmember utilities and individuals.

Projects are managed closely from their inception to the final report by the foundation's staff and large cadre of volunteers who willingly contribute their time and expertise. The foundation serves a planning and management function and awards contracts to other institutions such as water utilities, universities, and engineering firms. The funding for this research effort comes primarily from the Subscription Program, through which water utilities subscribe to the research program and make an annual payment proportionate to the volume of water they deliver and consultants and manufacturers subscribe based on their annual billings. The program offers a cost-effective and fair method for funding research in the public interest.

A broad spectrum of water supply issues is addressed by the foundation's research agenda: resources, treatment and operations, distribution and storage, water quality and analysis, toxicology, economics, and management. The ultimate purpose of the coordinated effort is to assist water suppliers to provide the highest possible quality of water economically and reliably. The true benefits are realized when the results are implemented at the utility level. The foundation's trustees are pleased to offer this publication as a contribution toward that end.

Edmund G. Archuleta, P.E.
Chair, Board of Trustees
Awwa Research Foundation

James F. Manwaring, P.E.
Executive Director
Awwa Research Foundation

## ACKNOWLEDGMENTS

The author of this report is indebted to the following individuals and water utility for their cooperation and participation in this project:

Dr. Nan Mei, Ms. Jane Lee, and Ms. Sharon Barker, Experimental Oncology, Cross Cancer Institute, Edmonton, Alberta, Canada

Drs. Hailin Wang, James Xing, Zhilong Gong, Guifeng Jiang, Mike Lam, Woei Tan, Christine Teixeira, and Mingsheng Ma, Ms. Xiufen Lu, Mr. Trevor Carnelley, Ms. Corinna Watt, Ms. Paula Murphy, Mr. Shengwen Shen, Ms. Sheila Jessa, Ms. Meiling Lu, and Ms. Robyn Kalke, Department of Public Health Sciences, University of Alberta, Edmonton, Alberta, Canada

Dr. William R. Cullen, Department of Chemistry, University of British Columbia, Vancouver, British Columbia, Canada

Dr. H. Vasken Aposhian, Department of Molecular and Cellular Biology, University of Arizona, Tucson, Arizona, USA

Dr. Les Gammie, EPCOR Water Services, Edmonton, Alberta, Canada

The authors wish to thank Ms. Katerina Carastathis for her assistance in preparing this report. The authors also thank Mr. Frank Blaha (Senior Project Manager), Dr. Jarka Popovicova (Project Manager) and Drs. Mariano Cebrian, Krista Clark, Mike Hughes, Curtis Klaassen, and Pankaj Parekh (Project Advisory Committee members) for their constructive suggestions and support for the project. The Awwa Research Foundation is acknowledged for its financial, technical, and administrative assistance in funding and managing the project.

## EXECUTIVE SUMMARY

Arsenic is naturally present in the environment. Chronic exposure to inorganic arsenic is a major concern in many parts of the world because of the elevated cancer risk associated with the ingestion of high levels of arsenic. The general aim of this study has been to extend our knowledge of the cellular response to exposure to arsenic. Much of our efforts have focused on damage to DNA and its repair. We have analyzed cells treated with arsenic alone and in combination with other genotoxic agents in order to identify mechanisms of DNA damage, the influence of arsenic on DNA repair and potential critical genes responsible for repair of arsenic induced DNA damage. To a limited extent, we have also addressed the important questions of arsenic metabolism and reaction with cellular protein. Whenever possible we have tried to use environmentally relevant doses of arsenic, i.e. < 300 μg/L.

While its mode of action has yet to be fully elucidated, oxidative DNA damage has been suggested as a possible mechanism. Employing a sensitive assay for thymine glycol (a common oxidative base lesion), we directly confirmed the induction of oxidative DNA damage in human cells exposed to sodium arsenite at environmentally relevant concentrations (<300 μg/L). A series of human cell lines associated with sensitivity to oxidative agents, including, Fanconi anemia, Bloom's syndrome, Xeroderma pigmentosum, Ataxia telangiectasia, Nijmegen breakage syndrome, M059J (cells lacking DNA-PK) and EM9 (*XRCC1*-mutated hamster cells) were then examined for their response to arsenic-induced cytotoxicity. These cell lines harbor mutations in genes involved in a variety of DNA repair pathways. Only the Ataxia telangiectasia cells displayed a marked hypersensitive response (>2-fold). The protective role of the ATM protein were confirmed by the normal response to arsenic displayed by Ataxia telangiectasia cells expressing wild-type ATM. Although the ATM protein plays a pivotal role in response to DNA double-strand break induction, none of the other cell lines with defects in double-strand break repair displayed a similar hypersensitivity. Further examination indicated that sodium arsenite at concentrations below 1 mg/L do not generate significant levels of double-strand breaks. Our data suggest that the ATM protein functions in an important but different capacity in the cellular response to arsenic toxicity than it does in response to agents that generate double-strand breaks, such as ionizing radiation. Furthermore, the lack of hypersensitivity to arsenic displayed by the other cell lines calls into question the hypothesis that DNA damage is a significant factor in arsenic cytotoxicity. It does not, however, rule out the possibility that arsenic acts as a co-carcinogen when cells are exposed to other DNA damaging agents.

Arsenic compounds have been shown to act synergistically with known carcinogens (e.g., UV light and alkylating agents) in the cytotoxicity and mutagenicity of animal cells, but little is known of the interaction between arsenic and γ-radiation. To address this question, we have examined the influence of sodium arsenite at the molecular level by monitoring the removal of thymine glycol, a radiation-induced DNA base lesion, and at the cellular level by measuring survival of cells exposed to combinations of ionizing radiation and arsenite. The assay for thymine glycol removal showed that arsenite slowed the rate of removal of the damaged base. In the survival assays, ten human cell lines were examined including three DNA repair-proficient cells A549 (human lung carcinoma cells), GM38 and GM43 (normal human fibroblasts), and seven DNA repair-deficient cells CRL-1223 (Xeroderma pigmentosum A), GM434 (XP-D), GM3021 (XP-G), FA1196 (Fanconi anemia), AG06040 (Bloom's syndrome), 780816 (NBS), and GM05823 (AT). The cells were irradiated to 2 Gy γ-radiation with pre- or post-incubation with various concentrations of arsenite, and exposed to 300 μg/L of arsenite with pre- or post-irradiation to different doses of γ-rays. Although the combination of the two agents was more toxic than either alone, the effect under all circumstances was additive not synergistic. This strongly implies that ionizing radiation and arsenic kill cells by distinct mechanisms.

Individuals exposed to arsenic will also be exposed to carcinogenic polycyclic aromatic hydrocarbons (PAH), such as benzo(a)pyrene, present in smoke and exhaust fumes. In order to study the influence of arsenic on the repair of PAH-DNA adducts, we established sensitive competitive and non-competitive immunoassays for the major benzo(a)pyrene adduct to DNA, i.e. the guanine adduct of benzo(a)pyrene diol epoxide (BPDE). The assays made use of capillary zone electrophoresis for separation, and laser-induced fluorescence for detection. The assays are relatively straightforward because there is no need for DNA digestion, reducing potential background due to extensive handling of DNA. In our preliminary study of the influence of arsenic on the generation and repair of the BPDE adducts we observed that the presence of arsenic significantly enhanced the level of the adducts but the kinetics of repair remained the same. This suggests that arsenic may either alter chromatin structure to render DNA more accessible to BPDE or inhibit other processes that inactivate BPDE, such as reaction with glutathione.

Another possible mechanism of arsenic toxic effects currently under extensive investigation deals with arsenic binding with proteins. However, few experimental data are available on arsenic-containing proteins in biological systems. No specific or unique arsenic binding protein has been characterized. A study on the interactions of arsenic with a model protein, metallothionein, is described here. Size exclusion chromatography with inductively coupled plasma mass spectrometry analysis of reaction mixtures between trivalent arsenicals and metallothionein clearly demonstrated the formation of complexes of arsenic with metallothionein. Analysis of the complexes using electrospray quadrupole time-of-flight tandem mass spectrometry revealed the detailed binding stoichiometry between arsenic and the 20 cysteine residues in the metallothionein molecule. Inorganic arsenite ($As^{III}$) and two trivalent methylation metabolites, monomethylarsonous acid ($MMA^{III}$) and dimethylarsinous acid ($DMA^{III}$), showed different binding stoichiometry with the metallothionein. Each metallothionein molecule could bind with up to 6 $As^{III}$, 10 $MMA^{III}$, and 20 $DMA^{III}$ molecules, consistent with the coordination chemistry of these arsenicals.

Biomethylation is the major human metabolic pathway for inorganic arsenic. Methylation of arsenic was previously thought to be a detoxification process because the pentavalent monomethylarsonic acid ($MMA^{V}$) and dimethylarsinic acid ($DMA^{V}$) are less toxic than the inorganic arsenic. However, recent findings of more toxic trivalent methylation metabolites, $MMA^{III}$ and $DMA^{III}$, suggest that biomethylation of arsenic may not be a detoxification process. To study the metabolism of arsenic, methods for speciation of these arsenic metabolites were developed._The methods involve ion pair chromatographic separation of arsenic species with hydride generation atomic fluorescence detection. Speciation of arsenate ($As^{V}$), arsenite ($As^{III}$), $MMA^{V}$, $MMA^{III}$, $DMA^{V}$, and $DMA^{III}$ in human urine samples was complete in 6 min. Detection limits were in the range of 0.5–2 μ g/L, allowing for direct urine analysis without any sample pretreatment.

# CHAPTER 1
# INTRODUCTION

Arsenic is classified as a human carcinogen. Epidemiological studies of populations exposed to high levels of arsenic due to ingestion of contaminated drinking water, including those from Taiwan (Chen et al. 1986, Chen et al. 1992, Chiou et al. 1995), Argentina (Hopenhayn-Rich et al. 1996a), and Chile (Smith et al. 1998), have suggested an association between the very high levels of arsenic and the prevalence of skin, lung and bladder cancers. Several reports from other areas of the world, such as West Bengal, India (Das et al. 1994, 1995; Chowdhury et al. 1997; Chatterjee et al. 1995), Inner Mongolia, China (Luo et al. 1997), Mexico (Cebrian et al. 1983), and Argentina (Astolfi et al. 1981) also attributed the cancer etiology to the ingestion of very high levels of arsenic from drinking water.

While ingestion of high levels of arsenic is believed to be a cause of certain cancers, estimates of cancer risk resulting from the exposure to low levels of arsenic are the subject of considerable debate (Smith et al. 1992; Carlson-Lynch, Beck, and Boardman 1994; Mushak and Crocetti 1995; Pontius, Brown, and Chen 1994; Chappell et al. 1997; Abernathy et al. 1999). There is a lack of an adequate animal model due to metabolic differences between rodents and humans with respect to arsenic. Assessment of cancer risk based on extrapolations from toxicological studies using rodents and from epidemiological studies involving highly exposed populations creates large uncertainties. Furthermore, the mechanism(s) of arsenic toxicity with respect to cancer are not clearly established (Chappell et al. 1997; Pontius, Brown, and Chen 1994; Yamauchi and Fowler 1994; NRC 1999; NRC 2001; Rossman 1998; Kitchin 2001; Hughes 2002). An improved understanding of arsenic toxicity and the dose-response relationship for relatively low levels of arsenic could improve the risk assessment process. The objective of this study was to contribute to a better understanding of arsenic health effects.

## ARSENIC

The toxicity and metabolism of arsenic differ dramatically depending on the chemical form of arsenic. A major process in arsenic metabolism is the methylation of arsenic. The process involves two-electron reduction followed by methyl group transfer (Figure 1.1) (Cullen, McBride, and Reglinski 1984; Thompson 1993; Styblo, Yamauchi, and Thomas 1995; Vahter 1983; Buchet and Lauwerys 1994). Methyltransferases are responsible for methyl transfer with S-adenosyl-methionine (SAM) as the methyl donor (Zakharyan et al. 1995, Aposhian 1997). Because an alteration of DNA methylation may lead to aberrant gene expression, recent studies have suggested that an alteration of DNA methylation status, either hypomethylation (Zhao et al. 1997) or hypermethylation (Mass and Wang 1997, Zhong and Mass 2001), may be possible mechanisms for carcinogenesis of arsenic. Zhao et al. (1997) proposed that arsenic methylation could result in the depletion of SAM, the same methyl donor for the methylation of DNA. This would lead to DNA hypomethylation. On the other hand, Mass and Wang (1997) found that exposure of human lung adenocarcinoma A549 cells to sodium arsenite (0.08-2 $\mu$M) produced significant hypermethylation within a 341-base pair fragment of the promoter of p53. However, these studies did not examine arsenic speciation although arsenic methylation was considered as an important process.

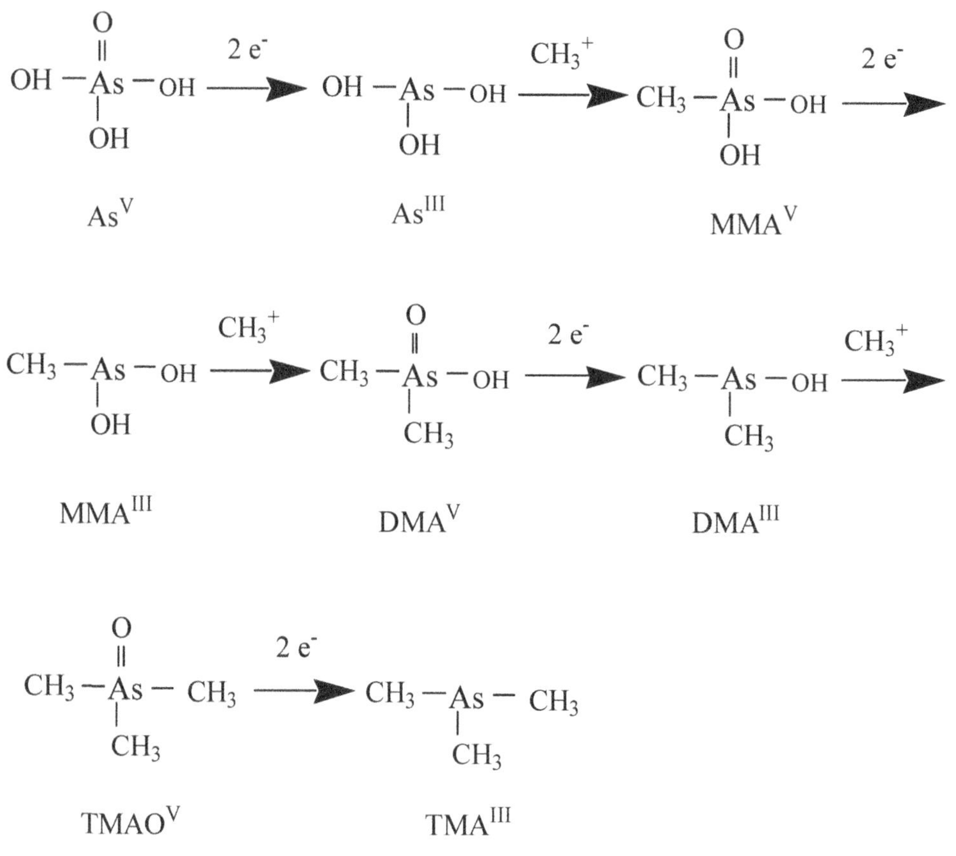

Figure 1.1. Biomethylation of arsenic showing a sequence of reduction of pentavalent to trivalent arsenic followed by oxidative addition of a methyl group. (*Source:* Le et al. 2000a)

Detailed speciation could contribute to a better understanding of arsenic metabolism and toxicity. We will describe speciation of arsenic that contributes to an improved understanding of arsenic metabolism (Chapter 6) and interaction with proteins (Chapter 7).

How chronic arsenic exposure causes cancer is still not clear, nor is the relationship between arsenic-induced cell injury/cell death and the carcinogenic response (Abernathy et al. 1999). Two principal ideas have emerged: induction of DNA damage and/or inhibition of DNA repair. There are some data to indicate that arsenic produces reactive oxygen species (Liu et al. 2001b) and may cause oxidative DNA damage (Matsui et al. 1999; Mei et al. 2002), but arsenic is not a potent mutagen in short-term tests (Rossman 1998). On the other hand, it does affect the mutagenicity of other carcinogens, probably via direct or indirect effects on DNA repair (Hartmann and Speit 1996; Hu, Su and Snow 1998). Arsenite has been shown to enhance the cytotoxicity, mutagenicity, and clastogenicity of ultraviolet (UV) light, alkylating and DNA crosslinking agents in rodent and human cells (Lee-Chen, Yu, and Jan 1992; Lee-Chen et al. 1993). Arsenite was found to potentiate UV killing of DNA repair-proficient normal human and Xeroderma pigmentosum (XP) variant fibroblasts but not that of DNA repair-defective XP group A cells (Okui and Fujiwara 1986). Arsenite has also been shown to inhibit the excision of pyrimidine dimers in normal human fibroblasts and in HeLa cells (Snyder, Davis, and Lachmann 1989). These synergistic effects were hypothesized to be due to arsenic inhibition of DNA repair enzymes. The inhibitory steps of arsenite in UV-irradiated rodent cells may differ from that in human cells.

Evidence for interactive effects of arsenic exposure in carcinogenesis can also be found in epidemiological studies. A synergistic interaction between arsenic exposure and cigarette smoking in the induction of lung cancer (Tsuda et al. 1995, Hertz-Picciotto et al. 1992) has been indicated. The biochemical mechanisms to explain such interactions have not been demonstrated but are consistent with a mechanism involving arsenic impairing DNA repair.

Chapters 2 and 3 of this report will address the issues of DNA damage, repair and cooperativity with another DNA damaging agent, in this case ionizing radiation.

## CELLULAR DEFENSE MECHANISMS

Two mechanisms by which to protect cellular DNA from the damaging effects of arsenic are (i) to prevent arsenic from causing the damage and (ii) to repair the DNA damage. Since arsenic reacts with protein sulfhydryl groups, metallothionein (MT) could be a potentially protective protein. Metallothionein is a low molecular weight (6000-7000 Da), cysteine-rich protein expressed in many cells. It has a high affinity for many metals and is known to effectively protect cells from cadmium toxicity. In recent studies, Park, Liu, and Klaassen (2001) showed that the $LD_{50}$ of arsenic for wild-type mice was 1.4-fold higher than for MT-null mice, and Liu et al. (2000) found that MT-null mice are more sensitive than wild-type mice to the hepatotoxic and nephrotoxic effects of chronic exposure to arsenic. Thus there is a clear need to further understand the interaction of MT with arsenic from the standpoint of cellular protection. At the same time, the interaction of arsenic with MT may serve as a model for the interaction of arsenic with other cellular proteins, including DNA repair proteins. Metallothionein-arsenic interactions form the topic of Chapter 7 of this report.

DNA repair is an essential process to protect against cancer. A human disorder, xeroderma pigmentosum (XP), is caused by defective repair of DNA damage induced by ultraviolet (UV) radiation. XP patients are extremely sensitive to sunlight and develop multiple cancers in areas exposed to sunlight (Cleaver and Kraemer 1989). Likewise, predisposition to a common form of colon cancer is due to a flaw in DNA mismatch repair (Modrich 1994, Modrich and Lahue 1996).

Natural genotoxic agents, man-made chemical carcinogens, as well as ionizing and UV radiation can damage DNA. If such damage were allowed to persist, human cells would cease to function properly, mutations would accumulate, and tumors would develop. Fortunately, all living organisms have DNA repair systems that cope with DNA damage, preserve genetic information, and guard against cancer.

The most common pathways for repairing DNA base damage are base excision repair (BER) and nucleotide excision repair (NER) (Sancar 1994). Base excision repair is a pathway commonly used to remove smaller modifications to DNA bases, such as is produced by oxidative processes e.g. thymine glycols and 8-oxoguanine, or methylation of DNA e.g. 3-methyladenine.

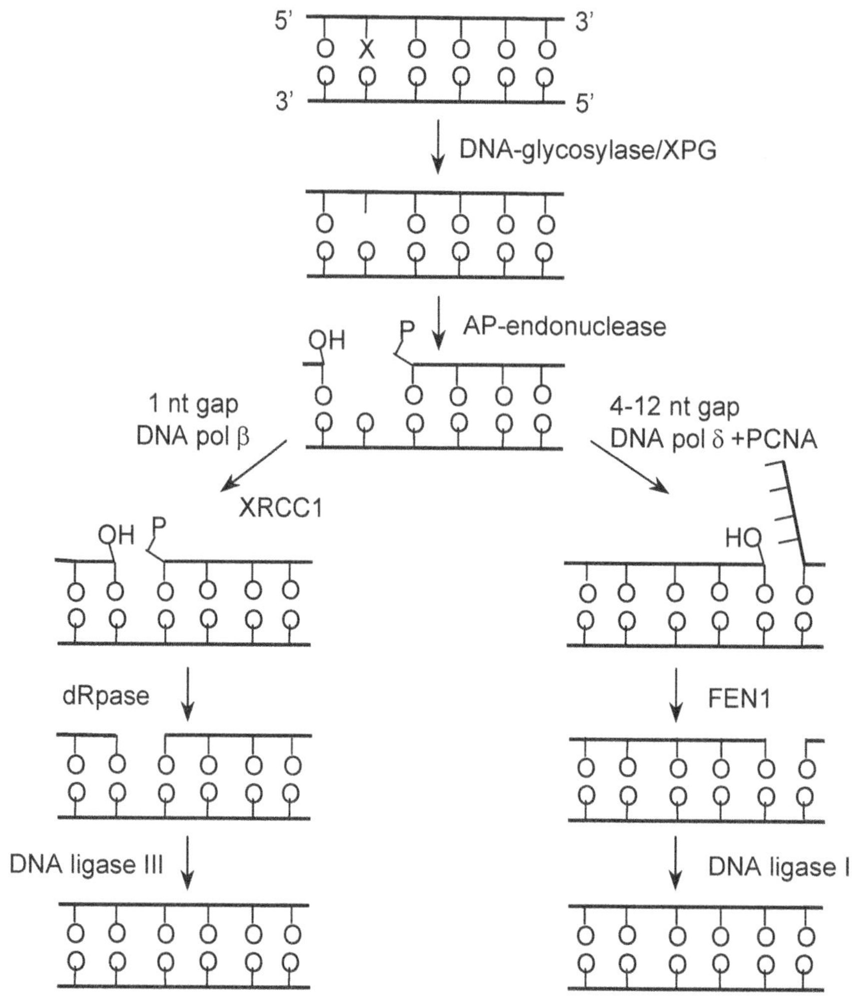


Figure 1.2. Schematic representation of base excision repair (BER). (*Source:* Adapted from Friedberg, Walker, and Siede 1995)

A schematic of the BER pathway is shown in Figure 1.2. The first step is the removal of the modified base by a DNA glycosylase to generate an abasic site (sometimes referred to as an apurinic/apyrimidinic or AP site). An AP endonuclease then cleaves the phosphodiester bond at the abasic site to produce a strand break with a deoxyribose phosphate still attached to the 5'-terminus. This is enzymatically removed by the deoxyribose phosphatase (dRpase) activity of DNA polymerase β and the single nucleotide gap is filled in by the polymerase function of DNA polymerase β. The strand is then rejoined by a DNA ligase. Each of the known DNA glycosylases appears to recognize a limited family of base modifications. For example the human hNth protein, like its *E. coli* homologue endonuclease III, recognizes several pyrimidine modifications including thymine glycol and dihydrothymine (Friedberg, Walker, and Siede 1995; Weinfeld, unpublished data). In mammalian cells, the major AP endonuclease appears to be a homologue of *E. coli* exonuclease III; in humans this enzyme is referred to as Ape or HAP1. DNA polymerase β and DNA ligase III are the mammalian enzymes that fulfill the remaining functions (Lindahl, Karran, and Wood 1997). Recently a second pathway subsequent to the AP endonuclease activity has been discovered in mammalian cells. Although less well characterized, it appears to require DNA polymerase δ in place of DNA polymerase β and DNA ligase I in place of DNA ligase III (Lindahl, Karran, and Wood 1997). In addition to the repair enzymes mentioned, there appears to be a role for the XPG protein in the glycosylase step (Klungland et al. 1999), and for XRCC1 in the processing and joining of strand break termini (Whitehouse et al. 2001).

Major distortions to the DNA, such as UV light-induced cyclobutane pyrimidine dimers and bulky adducts generated by polycyclic aromatic hydrocarbons, are repaired by the nucleotide excision repair (NER) pathway (Sancar 1994, 1996). This is a multienzymatic process involving some 30 gene products, including all those associated with the DNA repair disorder, xeroderma pigmentosum (XP). A schematic of the NER pathway is shown in Figure 1.3 (Batty and Wood 2000). In the human global NER pathway, the DNA lesion is first recognized by the XPC protein in association with hHR23B. This leads to the recruitment of replication protein A (RPA), XPA and the multisubunit transcription factor, TFIIH (including XPB and XPD), which helps to locally unwind the DNA. The lesion is then excised as part of an oligonucleotide approximately 30 bases in length. Excision 3' to the lesion is mediated by XPG and 5' to the lesion by a complex of XPF and ERCC1 proteins. The resulting single-strand gap is filled in by DNA polymerase δ or ε in association with other enzymes, and finally the strand is rejoined by a DNA ligase, most probably DNA ligase I (Lindahl, Karran, and Wood 1997). In a modification to this pathway, lesions in transcriptionally active DNA block RNA polymerase II, and the stalled complex is recognized by Cockayne's syndrome B (CSB) protein, which in turn recruits CSA protein and other members of the NER family (Hoeijmakers 2001).

Oxidative DNA damage often leads to the formation of single- and double-strand breaks. Single-strand break repair is considered to be a sub-pathway of BER involving XRCC1, DNA polymerase β and DNA ligase III (Whitehouse et al. 2001). Repair of double-strand breaks is considerably more complicated (reviewed by Hoeijmakers 2001). There are two principal pathways, non-homologous end-joining (NHEJ) and homologous recombination (HR). The former can be considered as a mechanism to force the two broken ends together even at the expense of losing some genetic information. The major proteins involved in this pathway include the Ku70/Ku80 dimer, DNA-dependent protein kinase catalytic subunit (DNA-$PK_{CS}$), XRCC4 and DNA ligase 4. HR requires a second copy of the damaged DNA, which can be provided by the second copy of each chromosome, and is considered to be error free. This pathway is not as well characterized and some of the proteins have not been identified. Proteins known to be involved in this pathway include the rad50/Mre11/NBS complex, rad51, rad52 and rad54. NBS stands for Nijmegen breakage syndrome, which is a rare genetic syndrome with neurological defects similar to Ataxia telangiectasia (AT). The protein responsible for AT, the ATM protein (ataxia telangiectasia mutated), itself plays a key role in both pathways by acting as a regulator in response to DNA double-strand breaks and integrating repair with cell cycle check points (Bakkenist and Kastan 2003). Cells derived from AT and NBS patients and those with defects in the Ku or DNA-$PK_{CS}$ are highly sensitive to ionizing radiation.

Chapters 2, 3 and 5 detail our studies of the influence of arsenic on DNA damage and repair in mammalian cells, either in isolation or in combination with other genotoxic agents. Whenever possible we have tried to use environmentally relevant doses of arsenic, i.e. $< 300$ μg/L.

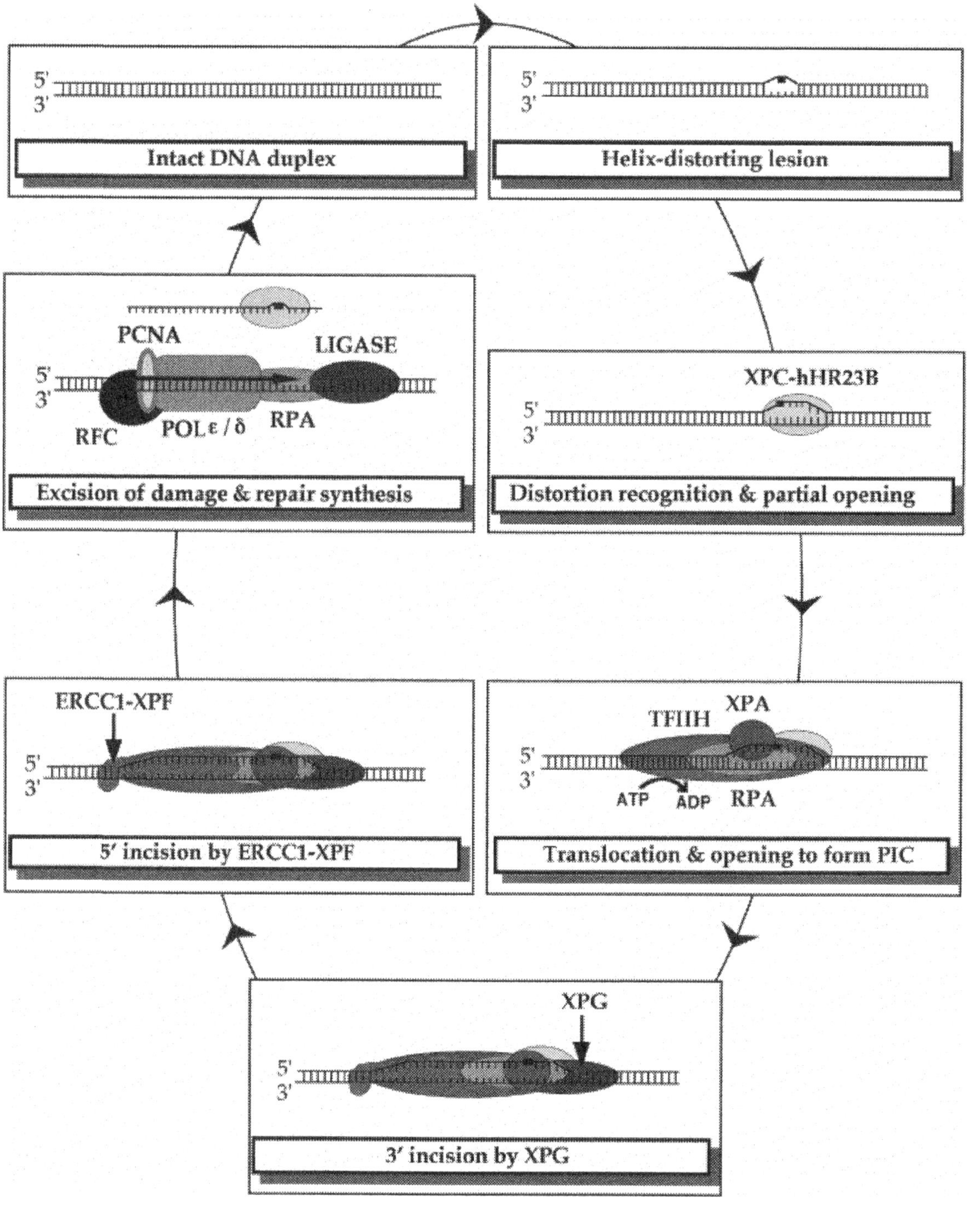


Figure 1.3. Model for nucleotide excision repair (NER) in mammalian cells showing recognition of damage, opening of a region around the lesion, dual incision, and repair synthesis. (*Source:* Batty and Wood 2000)

## ASSAYS FOR DNA DAMAGE

Several studies have examined arsenic effects on DNA repair, but they have used much higher levels of DNA damaging agents than those commonly encountered by humans. For instance, Lee-Chen et al. (1993) used as high as 1000 μM methyl methanesulfonate to cause DNA damage to CHO cells, and then monitored the repair of DNA double-strand breaks in the presence of 5-160 μM (375-12,000 μg/L) of arsenic. Although these studies are useful for providing preliminary information, extrapolation from high doses used in these DNA repair studies to the environmentally relevant doses carries large uncertainties. The same problem occurs looking at repair of DNA caused by radiation and other environmental toxicants and clinically used agents. The main reason for using much higher than clinically and environmentally relevant doses of DNA damaging agents has been a lack of sensitive assays for detecting low levels of DNA base damage. Although pulsed-field gel electrophoresis and single-cell gel electrophoresis (the comet assay), have the capacity to detect damage at low doses of DNA damaging agents, they are mainly restricted to the measurement of strand breaks (Longo et al. 1997; Fairbairn, Olive, and O'Neill 1995).

Highly sensitive assays for DNA damage are required in order to study the effects of arsenic on DNA damage and repair. Many techniques have been developed for the determination of DNA damage (Pfeifer 1996). The most commonly used techniques include $^{32}$P-postlabeling assays (Randerath, Reddy, and Gupta 1981; Beach and Gupta 1992; Cadet et al. 1992; Randerath and Randerath 1994; Phillips 1997), single cell gel electrophoresis (comet assay) (Singh et al. 1988; Singh, Stephens, and Schneider 1994; Fairbairn, Olive, and O'Neill 1995), gas chromatography with mass spectrometry (GC/MS) detection (Fuciarelli et al. 1989, Annan et al. 1989, McCloskey and Crain 1992, Chiarelli and Lay Jr. 1992, Giese 1997), and immunoassays (Levine et al. 1966; Leadon 1988; West, West, and Ward 1982; Poirier 1981; Santella et al. 1988; Mitchell 1996; Thomale et al. 1996; Melamede et al. 1996; Le et al. 1998; Xing et al. 2001). These assays each have their advantages and disadvantages with respect to sensitivity, specificity, and background levels. The most sensitive method for detecting DNA damage is the radioactive $^{32}$P-postlabeling method, which allows the detection of one adduct in $10^9$ unmodified nucleotides in microgram amounts of DNA (Randerath, Reddy, and Gupta 1981; Beach and Gupta 1992; Cadet et al. 1992; Randerath and Randerath 1994; Phillips 1997). This technique requires working with hazardous radioactive material and time-consuming, multiple chromatographic separation procedures. In addition, $^{32}$P-postlabeling methods in general do not provide information on adduct identity. The single cell gel electrophoresis, or comet assay, is very sensitive, but it is primarily for the analysis of DNA strand breaks (Singh et al. 1988; Singh, Stephens, and Schneider 1994; Fairbairn, Olive, and O'Neill 1995). GC/MS methods require enzymatic digestion and chemical derivatization of the DNA, leading to potential artifacts because the extensive DNA treatment procedures can introduce oxidative damage to the DNA. Electrochemical techniques are limited to the detection of modified bases that are electrochemically active (Floyd et al. 1986, Shigenaga and Ames 1991, Wang et al. 1997).

Recent advances in mass spectrometry have led to a number of reports making use of mass spectrometry detection with high performance liquid chromatography (HPLC) (Wolf and Vouros 1994; Lim et al. 1997; Huang et al. 1998a; Muller et al. 1997; Hakala et al. 1999; Roberts et al. 2001; Andrews, Vouros, and Harsch 1999), capillary electrophoresis (CE) (Andrews, Vouros, and Harsch 1999; Barry, Norwood, and Vouros 1996; Deforce et al. 1996), and capillary electrochromatography (CEC) separation (Ding and Vouros 1997; Ding et al. 1998; Gangl, Turesky, and Vouros 1999). Various fluorescence techniques, such as fluorescence line narrowing, synchronous fluorescence, and laser-induced fluorescence, have also been studied for DNA damage analysis (Jankowiak et al. 1988; Duhachek et al. 2000; Weston and Bowman 1991; Li, Hurtubise, and Weston 1999; Rogers et al. 1999). Several techniques are based on the polymerase chain reaction (PCR) (Pfeifer et al. 1991; Sano, Smith, and Cantor 1992; Jennerwein and Eastman 1991; Kalinowski, Illenye, and Van Houten 1992; Govan, Valles-Ayoub, and Braun 1990; Zhang and Poirier 1997), which allow for the identification of DNA damage at specific locations in the genome. While these new developments are useful for measuring relatively high levels of DNA damage, a major challenge remains to be the high sensitivity required for detecting low levels of DNA damage induced by environmentally relevant exposure to DNA damaging agents.

We have developed assays for detection of trace levels of DNA modifications. They make use of antibodies to bind DNA specific lesions, capillary zone electrophoresis (CE) for separation, and post column laser-induced fluorescence (LIF) for detection. We will describe these assays in Chapters 2-5 and their use to study the effects of arsenic on cellular DNA when cells are exposed to arsenic alone or in combination with $\gamma$ radiation (Chapters 2 and 3) or benzo[a]pyrene and its major metabolite (Chapters 4 and 5).

# CHAPTER 2
# GENETIC PREDISPOSITION TO THE CYTOTOXICITY OF ARSENIC: THE ROLE OF ATM

## INTRODUCTION

Arsenic is widely distributed in our environment. Exposure to arsenic, mostly via drinking water, has been associated with cancer of the skin and various internal organs (Chen et al. 1992, Smith et al. 1998, NRC 1999). The current maximum contaminant level (MCL) of arsenic allowed in drinking water in the USA is 10 μg/L. In Canada, the maximum allowable limit of arsenic in drinking water is 25 μg/L. However, in some areas of the world the natural arsenic level can be as high as several thousand μg/L (Bagla and Kaiser 1996), and in many countries, including India (Bagla and Kaiser 1996) and Bangladesh (Nickson et al. 1998), such high arsenic levels constitute a serious public health concern. The increase in cancer risk is attributed mainly to the presence of inorganic arsenic. Ironically, inorganic arsenic compounds have also been found to have beneficial uses. Indeed, several arsenic compounds are undergoing testing for clinical efficacy in cancer treatment, in particular arsenic trioxide, which has been used as an extremely specific and successful chemotherapeutic for acute promyelocytic leukemia (Chen et al. 1997).

How chronic arsenic exposure causes cancer is still not clear. Nor is the relationship between arsenic-induced cell injury/cell death and the carcinogenic response (Abernathy et al. 1999). Attention is being focused on the induction of DNA damage and/or inhibition of DNA repair. There are data to indicate that arsenic may cause oxidative DNA damage. Several groups have reported the clastogenic effects of arsenical compounds, though this is not necessarily proof of oxidative damage (Mass et al. 2001). Recently, Lynn et al. (2000) and Li et al. (2001a) observed an increase in arsenic-induced strand cleavage after incubation of the DNA with *Escherichia coli* formamidopyrimidine-DNA glycosylase, an enzyme that cleaves DNA at oxidized guanines, and Matsui et al. (1999) reported high levels of 8-oxoguanine in the DNA of patients with arsenic-related Bowen's disease. For this study, we have made use of a sensitive immunoassay to analyze the induction of thymine glycol (Le et al. 1998), a well-characterized DNA base lesion generated by ionizing radiation and other oxidative processes (Figure 2.1). We report that cellular exposure to low doses of arsenic (<300 μg/L) does indeed generate thymine glycol, thus providing strong direct evidence for arsenic-induced oxidative DNA damage.

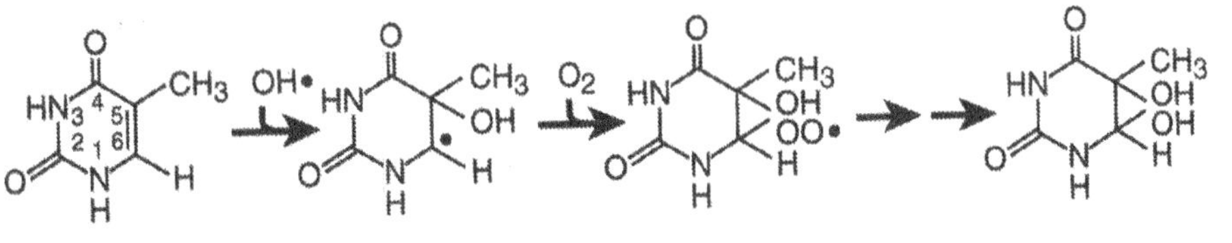

Figure 2.1 Oxidation of thymine to thymine glycol (*Source:* Mei et al. 2003)

This line of inquiry was followed up by asking whether certain mammalian cell lines, associated with sensitivity to oxidative damage and/or DNA repair deficiency, display an increased cytotoxic response to arsenic. The cells under consideration included those isolated from individuals with Xeroderma pigmentosum (XP, complementation groups, A, D and G), Fanconi anemia (FA), Bloom's syndrome (BS), Nijmegen breakage syndrome (NBS), Ataxia telangiectasia (AT), as well as a human cell line deficient in DNA-dependent protein kinase and hamster cell lines with mutations in *XRCC1*, *XRCC2* or *XRCC3*. We found that of these, only AT cells display a marked sensitivity to arsenic. The product of the ATM (ataxia telangiectasia mutated) gene is a key player in the cellular response to DNA double-strand break formation (Rotman and Shiloh 1999, Bakkenist and Kastan 2003). However, we subsequently observed that sodium arsenite is not a potent inducer of double-strand breaks. This observation, taken together with other cell survival data, suggests that double-strand breaks are not responsible for arsenic-induced cytotoxicity, and that the ATM protein plays a role in preventing As toxicity other than responding to double-strand break formation.

## MATERIALS AND METHODS

### Cell Lines and Culture Conditions

A549 (human lung carcinoma cells), CRL-1223 (XP complementation group A fibroblasts, XP-A), CRL-1196 (FA fibroblasts) and AT2BE (AT fibroblasts) were obtained from the American Type Culture Collection (Rockville, Md.). GM43 (normal human fibroblasts), GM0637A (SV40-transformed normal human fibroblasts), GM434 (XP-D fibroblasts), GM3021 (XP-G fibroblasts), AG06040 (BS fibroblasts), AT5BI (GM05823 AT fibroblasts) and GM05849C (SV40-transformed GM05823 AT fibroblasts) were supplied by the NIGMS Cell Repository (Camden, N.J.). Nijmegen breakage syndrome fibroblasts (780816 cells) were kindly provided by Dr. P. Concannon (Virginia Mason Research Center, Seattle, Wash.). M059J and M059K malignant glioma cells were kindly provided by Dr. J. Allalunis-Turner (Cross Cancer Institute, Edmonton, Alta.). The AA8, EM9, irs1SF, V79 and irs1 hamster cells were kindly provided by Dr. Larry Thompson (Lawrence Livermore National Laboratory, Livermore, Calif.) and Dr. David Murray (Cross Cancer Institute, Edmonton, Alta.). These cell lines were cultured in Dulbecco's modified Eagle's medium/nutrient mixture F-12 (D-MEM/F-12) (1:1 ratio) supplemented with 10% fetal bovine serum, penicillin (50 U/ml), streptomycin (50 mg/ml), L-glutamine (2 mM), non-essential amino acids (0.1 mM) and sodium pyruvate (1 mM), and maintained at 37°C in humidified incubators containing 5% $CO_2$. Unless otherwise noted, all culture supplies were purchased from Gibco BRL Life Technologies (Rockville, Md.).

The SV40-transformed AT cell lines, FT/pEBS7 cells, which carry the mammalian expression vector pEBS7 with a hygromycin resistance marker, and FT/pEBS7-YZ5 cells, which carry the vector containing the complete cDNA for the ATM gene (Ziv et al. 1997) were obtained from Dr. T. Jorgensen (Georgetown University, Washington, D.C.) and Dr. J. Larmer (University of Virginia, Charlottesville, Va.) from original stocks provided by Dr. Y. Shiloh (Tel Aviv University, Israel). FT/pEBS7 and FT/pEBS7-YZ5 cells were grown in D-MEM/F-12 (1:1 ratio) and 100 μg/ml hygromycin B (Roche Diagnostics GmbH, Germany) supplemented with 15% fetal bovine serum, penicillin (50 U/ml), streptomycin (50 mg/ml), L-glutamine (2 mM), non-essential amino acids (0.1 mM) and sodium pyruvate (1 mM), and maintained at 37°C in humidified incubators containing 5% $CO_2$.

## Treatment with Sodium Arsenite

Sodium arsenite [As(III)] was obtained from Sigma-Aldrich (Milwaukee, Wis.). The stock solution (1000 mg of As/L) was prepared in double-distilled water and sterilized by passing through a 0.22-μm (pore size) syringe filter. Arsenic concentration in the stock solutions was standardized against an arsenic atomic absorption standard solution (Sigma-Aldrich) using both inductively coupled plasma mass spectrometry (VG Elemental/TJL Solutions, Franklin, Mass.) and flame atomic absorption spectrometry (Le and Ma 1998). Working concentrations containing less than 10 mg of As/L were prepared fresh daily by diluting the stocks with D-MEM/F12 medium. The cells were treated with different arsenite concentrations from 25-300 μg/L for 24 hours. After treatment, the cells were either harvested for measurement of DNA damage or washed with phosphate-buffered saline (PBS), and 4 ml fresh normal medium were added, then the cells were incubated as described above for the colony-forming assay.

## Cell Irradiation

Cells in cultures were irradiated to various γ-ray doses (1-5 Gy) using a Shepherd Mark I-68A $^{137}Cs$ Irradiator (J. L. Shepherd & Associates, San Fernando, Calif.) at a dose rate of 1.23 Gy/min. Control cells were subjected to similar treatment but without irradiation.

## Colony-Forming Assay

To determine the cytotoxic response of different cell lines to arsenic or γ-radiation, exponentially growing cells were trypsinized, and resuspended in D-MEM/F12 medium. The cells were seeded into 60-mm diameter tissue culture dishes at densities from 50 to 10,000 cells per dish on the basis of preliminary experiments, allowed to attach for a period of 16-24 hours, then treated with arsenic or irradiated. To maintain the active growth of cells, 2 ml of fresh medium was added to each dish after 7 days. Cultures were incubated for 2 weeks in a humidified atmosphere of 5% $CO_2$ at 37°C before staining with 0.25% methylene blue. Surviving fractions were calculated by dividing the number of colonies in treated dishes by the number of colonies arising in sham-treated controls. To compare the sensitivity of the cell lines to appropriate controls, the survival curves were analyzed by the method of Thames and Rasmussen (1978).

## Measurement of Thymine Glycol

Details of the protocols used have been described by Le et al. (1998) and Xing et al. (2000). Briefly, the cells were lysed and DNA purified immediately after treatment with As(III) using DNAzol genomic DNA isolation solutions (as described by Gibco-BRL). The DNA concentrations were determined by UV spectrophotometry. To assay for 5,6-dihydroxy-5,6-dihydrothymine (thymine glycol, Tg), a primary mouse monoclonal anti-Tg antibody was used to selectively recognize the DNA base lesion (kindly provided by Dr. S.A. Leadon, University of

North Carolina School of Medicine), and an Alexa Fluor® 546 F(ab')$_2$ fragment of goat anti-mouse IgG (H+L), (Molecular Probes, Eugene, Ore.) was used as a fluorescent probe of the primary antibody. Capillary electrophoresis was used to separate the complex of antibody (primary and secondary) bound to DNA from unbound antibody, and laser-induced fluorescence detection was used to measure the fluorescent species.

**Measurement of DNA Double-Strand Breaks (Phosphorylated Histone H2AX)**

We used an approach similar to published methods (Paull et al. 2000, Sedelnikova et al. 2002). GM43 cells were cultured on cover slips placed in 35-mm dishes. Cells were exposed to 1 mg/L sodium arsenite for 24 hours at 37°C, after which time the cells were rapidly washed three times with PBS and the cells incubated in fresh media for 30 minutes to allow time for phosphorylation of histone H2AX. Other GM43 cells were irradiated with 0, 1 or 2 Gy and then incubated for a further 30 min. The media was removed and the cells were washed with PBS and then held in methanol/PBS (50:50 v/v) at room temperature for 10 min before fixing them in methanol at -30°C for 30 min. After removal of the methanol, the cells were held in PBS at room temperature for 5 min and then blocked with 5% milk powder in PBS for 30 min. The cells were then stained with anti-phospho-histone H2AX antibodies (Upstate Biotechnology, Lake Placid, N.Y.) for 1.5 h at room temperature in the dark. After removal of the primary antibody, the cells were washed with PBS, followed by 0.1% Tween 20 in PBS and three more times with PBS, and then incubated with Alexa Fluor 488-labeled goat anti-mouse IgG antibody (Molecular Probes, Eugene, Ore.) at room temperature in the dark for 45 min. The cells were again washed with PBS, followed by 0.1% Tween 20 in PBS and three more times with PBS and rinsed with water. The cover slips were mounted on microscope slides and stored at 4°C in the dark.

The fluorescent foci were detected and analyzed by confocal microscopy using a Zeiss LSM 510 microscope (Zeiss Instruments, Jena, Germany) and the Imaris Software package (Bitplane Inc., Zurich, Switzerland).

**Western Blotting Assay**

Cells were lysed in RIPA buffer (150 mM NaCl, 1% Nonidet P-40, 0.5% sodium deoxycholate, 0.1% SDS, 50 mM Tris, pH 7.5), containing a freshly added cocktail of protease inhibitors (Sigma), on ice for 30 min with occasional rocking. The lysates were centrifuged at 10,000×g for 15 min and the supernatant was collected. Protein samples (20 μg each) were resolved by SDS-polyacrylamide gel electrophoresis and then transferred to nitrocellulose membranes (Bio-Rad, Mississauga, Ont.) using the SemiDry transfer System (Bio-Rad). After blockage with 5% nonfat-milk in PBS and 0.01% Tween 20 for 1 h at room temperature, the membranes were incubated with antibodies specific to p53 (DO-1) and actin (I-19) from Santa Cruz Biotechology (Santa Cruz, Calif.), Ser15-phosphorylated p53 (Cell Signaling Technology, Beverly, Mass.), or ATM (NB 100-104, Novus Biologicals, Littleton, Colo.), for 1 h. Membranes were washed, blotted with secondary antibody conjugated with horseradish peroxidase (Jackson ImmunoResearch, West Grove, Pa.), followed by detection by enhanced chemiluminescence (Amersham Pharmacia, Baie d'Urfé, P.Q.).

**Cell Cycle Analysis**

Exponentially growing cells were cultured in the presence of As(III) for 24 h, fixed in ethanol at –20 °C and treated with propidium iodide (5 μg/ml) and RNase (0.1 mg/ml) at room temperature for 1 h. DNA content in $2 \times 10^5$ cells was determined using a FACSort flow cytometer (Becton Dickinson, San Jose, Calif.) equipped with CellQuest and ModFit software (Becton Dickinson).

## RESULTS

### Induction of Thymine Glycol (Tg) by Sodium Arsenite

We made use of a recently developed immunoassay to measure Tg (Le et al. 1998). The assay takes advantage of the high sensitivity afforded by laser induced fluorescence detection coupled with capillary electrophoresis separation, which allowed us to measure Tg after treatment of human A549 cells with a dose of sodium arsenite as low as 50 ppb (equivalent to 50 μg/L or 0.7 μM). As shown in Figure 2.2, there was a clear dose-response. The plot indicates that a 24-h exposure to 50-300 μg/L $As^{III}$ generates ~3-30 Tg per $10^8$ DNA bases.

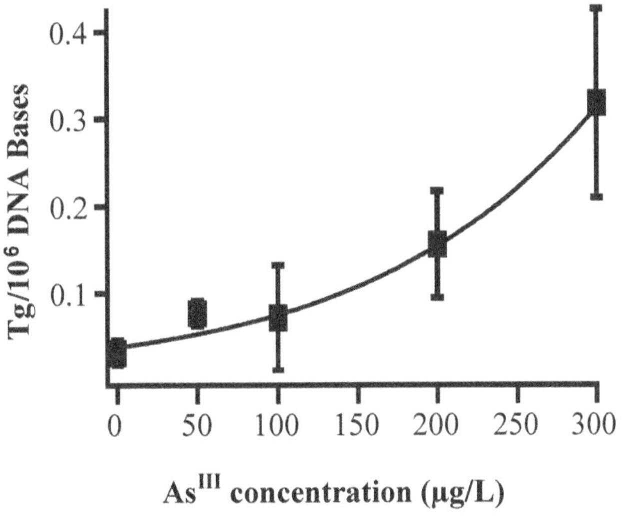


Figure 2.2 Generation of Tg in the DNA of human A549 cells by cellular exposure to sodium arsenite for 24 h. The lesion was measured by an immunoassay using capillary electrophoresis and laser-induced fluorescence to separate and detect the fluorescent species as described in Materials and Methods. Means are plotted for three separate determinations (±SE). The curve was generated by a best-fit polynomial. (*Source:* Mei et al. 2003)

## Dose and Time Dependence of Sodium Arsenite-Induced Cytotoxicity of A549 Human Cells

To determine the dose-response of A549 cells to $As^{III}$, exponentially growing cells were treated with $As^{III}$ for 4 h, 24 h or 6 days using different low-level doses from 25 to 300 μg/L. Arsenite induced dose-dependent toxicity in A549 cells as shown in Figure 2.3a. The mean plating efficiency of untreated A549 cells was 76%. The survival data fit well to a semi-log plot, and the lack of a shoulder in the dose response curves is consistent with previous reports that arsenic inhibits DNA repair or recovery processes that normally act to produce a shoulder (Hei, Liu, and Waldren 1998). The cell killing in A549 increased with treatment time. Figure 2.3b shows the effect of the time of arsenite treatment on the mean lethal doses, $D_{37}$, defined as the concentration that reduced survival to $e^{-1}$ (~ 0.37) in the log-linear portion of the curves for different treatment. It also appears that $D_{37}$ decreased quickly between 4 and 24 hours and more slowly afterwards. The results presented here agree with the findings of Hartwig et al. (1997), who reported the uptake of $As^{III}$, as well as the intracellular arsenic concentration in human fibroblasts, increased in a time-dependent manner and reached a maximum after 18 hours incubation.

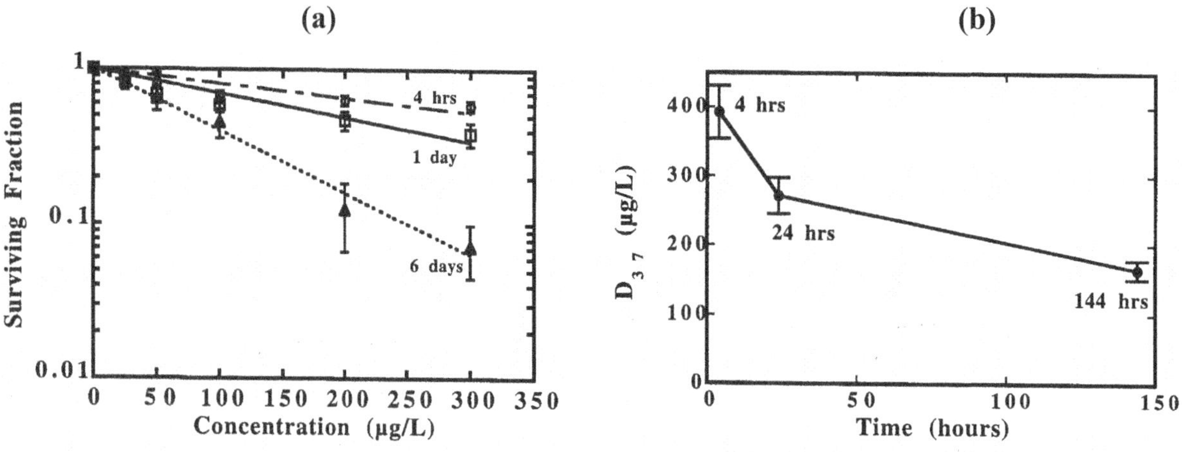


Figure 2.3 Cytotoxicity of sodium arsenite towards the A549 human lung carcinoma cell line. (a) Exponentially growing cells were treated with increasing doses of arsenite for 4 h, 24 h and 6 days. Each data point represents the mean of four experiments (±SD). (b) The effect of time of arsenite treatment on the mean lethal dose ($D_{37}$). (*Source:* Mei et al. 2003)

## Comparison of the Cytotoxic Effects of Four Arsenic Compounds

We also compared the dose-response of A549 cells to sodium arsenate ($As^V$), monomethylarsonic acid ($MMA^V$) and dimethylarsinic acid ($DMA^V$) to sodium arsenite ($As^{III}$) in similar experiments using the same dose range. Figure 2.4 shows the results of the colony forming assays of A549 cells and two other cell lines, GM43 (normal human fibroblasts) and CRL-1223 (fibroblast derived from an individual with the DNA repair disorder, Xeroderma pigmentosum), treated with the four arsenic compounds for 24 hours. The cells were not particularly sensitive to low doses of $MMA^{V}$ or $DMA^{V}$. $As^{V}$ was about 2 fold less toxic than $As^{III}$. Others have also reported that several different types of human cells have a sensitivity to $As^{V}$ approximately 3 to 10 fold less than $As^{III}$ using neutral red or tetrazolium dye uptake assay (Hu, Su, and Snow 1998).

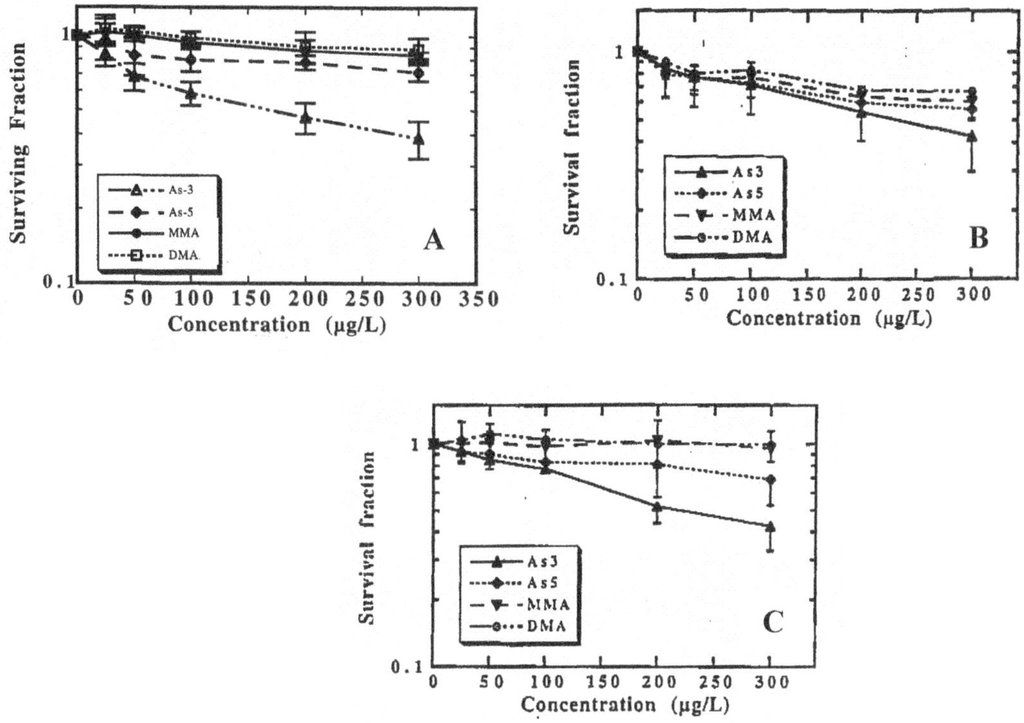


Figure 2.4 Comparison of the cytotoxic effects of 24-h exposure to increasing doses of sodium arsenite (As3), sodium arsenate (As5), monomethylarsonic acid (MMA) and dimethylarsinic acid (DMA) with A549 cells (A), GM43 cells (B) and CRL-1223 cells (C). Data points represent the mean of at least three experiments (± S.D.). (*Source:* Mei et al. 2003)

As a result of these and the previous preliminary studies, sodium arsenite was chosen as the arsenic compound for further study and doses and time of exposure for human cells was set at 25-300 µg/L and 24 h, respectively.

**Cytotoxic Response of DNA Repair Deficient Cell Lines Towards Sodium Arsenite**

The observation that arsenic induces oxidative damage led us to ask if cells regarded as sensitive to oxidative damage would show an elevated cytotoxic response to sodium arsenite. Accordingly a series of cell lines was subjected to the colony-forming assay for cell survival following exposure to various doses of sodium arsenite. Included were EM9 cells, which are Chinese hamster ovary cells harboring a mutation in the DNA base excision repair/single-strand break repair gene *XRCC1*, and representative cell lines from the human autosomal recessive disorders, Xeroderma pigmentosum (XP), Fanconi anemia (FA), Bloom's syndrome (BS) and Ataxia telangiectasia (AT). The clonogenic survival curves for the Chinese hamster ovary cells following 24-h exposure to 0-3000 μg/L $As^{III}$ are shown in Figure 2.5a. [The doses of $As^{III}$ used with these cells were greater than used subsequently with human cells because rodent cells are less sensitive to $As^{III}$ than human cells (NRC 1999)]. EM9 cells were shown to have similar sensitivity to the parental AA8 cells.

The responses of the FA, BS and XP cell lines following 24-h exposure to 0-300 μg/L sodium arsenite are shown in Figure 2.5b. Also included in this plot are the survival curves for two DNA-repair proficient human cell lines, the lung carcinoma cell line A549, and the normal fibroblast line, GM43. The results indicate a fairly narrow range of sensitivities with most cells falling between 40 and 70% survival at 300 μg/L. None of the FA, BS or XP cell lines showed any remarkable response. The most sensitive, AG06040, the BS line, had a very similar response to the repair-proficient cell line, GM43.

**Response of AT Cells to Sodium Arsenite**

We initially examined two AT fibroblast lines, AT2BE and AT5BI. Both gave very similar responses (Figure 2.5C) and were observed to be significantly more sensitive than the other human cell lines examined. The dose of arsenic required to reduce cell survival to 50% was >2.5-fold greater for GM43 cells than either of the AT cell lines, and the surviving fraction of the AT cells at 300 μg/L $As^{III}$ was only 0.15 compared with 0.4 for GM43. Further support for the ATM involvement in the cellular response to arsenic was provided by the use of ATM-complemented cells. The FT/pEBS7-YZ5 cell line was generated by Ziv et al (1997) by complementation of AT22IJE-T cells, an SV40-transformed AT cell line, with the complete cDNA for the ATM gene carried in the pEBS7 mammalian expression vector. FT/pEBS7, which is the control AT cell line bearing the empty vector, is more sensitive than GM43 to radiation (Figure 2.5D), but less sensitive than the untransformed AT cell lines. A reduction in sensitivity to radiation due to SV40 transformation has been reported by others (Jorgensen, Russell, and McRae 1995), and we observed that FT/pEBS7 cells have an almost identical radiation-survival response as GM05849C, which is the SV40-transformed AT5BI cell line. Interestingly FT/pEBS7 cells displayed a similar intermediate response to sodium arsenite (Figure 2.5D). The FT/pEBS7-YZ5 cells, on the other hand, were observed to have a normal survival response to both radiation and arsenite, which indicates that complementation with ATM restores resistance to radiation and arsenite. Confirmation that the FT/pEBS7-YZ5 cells expressed ATM protein is shown in the western blot in Figure 2.5E.

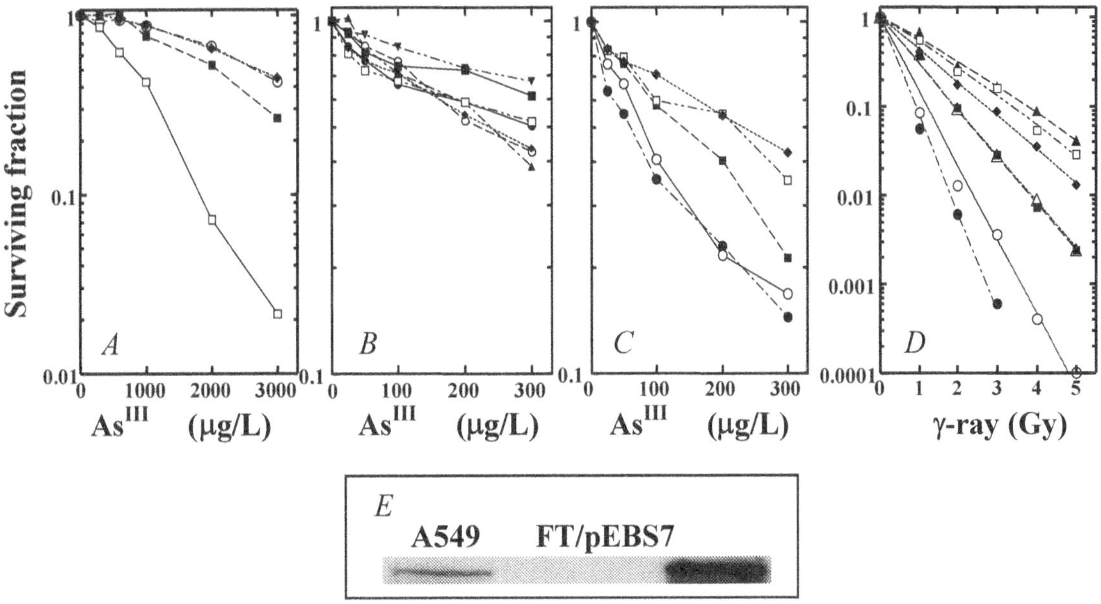


Figure 2.5 (A) Response of Chinese hamster ovary cells to 24-h exposure to sodium arsenite. Standard colony forming assay was used to examine the survival of the following cells. (■), AA8 (wild type); (□), EM9 (the *XRCC1* mutant). (B) Clonogenic survival of human fibroblasts following 24-h exposure to increasing concentrations of $As^{III}$. The data show the results of standard colony forming assays and are drawn from 4~8 determinations for each cell line. The following cells were examined (■) A549 (DNA-repair proficient human lung carcinoma), (◆) GM43 (normal human fibroblasts), (O) CRL-1223 (XPA fibroblasts), (□) GM434 (XPD fibroblasts), (●) GM3021 (XPG fibroblasts), (▼) FA1196 (FA fibroblasts) and (▲) AG06040 (BS fibroblasts). (C) and (D) Response of Ataxia telangiectasia cells to 24-h exposure to sodium arsenite (C) and γ-radiation (D). Each data point represents the mean of 4-6 determinations. The cells examined include (◆) GM43 (normal human fibroblasts), (O) AT2BE (AT fibroblasts), (●) AT5BI (AT fibroblasts), (■) FT/pEBS7 (SV40-transformed AT fibroblasts transfected with the vector pEBS7), (□) FT/pEBS7-YZ5 (SV40-transformed AT fibroblasts complemented with ATM cDNA), (Δ) GM5849 (SV40-transformed AT5BI fibroblasts) and (▲) GM0637A (SV40-transformed normal fibroblasts). (E) Western blot analysis of ATM protein expression in the indicated cell lines. (*Source:* Mei et al. 2003)

**Response of Double-Strand Break Repair Deficient Cells to Sodium Arsenite**

MO59J cells lack active DNA-dependent protein kinase, which plays an integral role in non-homologous end-joining, and consequently display a defect in double-strand break repair (Lees-Miller et al. 1995, Lieber 1999). In agreement with previous reports (Allalunis-Turner et al. 1995, Weinfeld et al. 1997), MO59J cells display a marked radiation hypersensitivity (Figure 2.6B), but do not appear to be sensitive to sodium arsenite in comparison to GM43 (Figure 2.6A), although they are slightly more sensitive at higher arsenic doses than DNA-PK positive MO59K cells, which were isolated from the same glioblastoma (Lees-Miller et al. 1995, Allalunis-Turner et al. 1995).

Cells derived from individuals with Nijmegen breakage syndrome (NBS) show a marked radiosensitivity. The NBS1 protein participates in a complex with Mre11 and Rad50 in homology-driven double-strand break repair (Carney et al. 1998). It has recently been shown by several groups that the NBS1 is a substrate for phosphorylation by ATM in response to ionizing radiation (Lim et al. 2000, Gatei et al. 2000, Zhao et al. 2000). We therefore examined the sensitivity of NBS cells to arsenic. Figure 2.6b confirms the radiosensitivity of the NBS cell line used, but Figure 2.6a reveals that NBS cells display a very modest increased sensitivity to arsenic (~1.4-fold) compared to GM43 cells at an equitoxic dose and a considerably smaller difference at the same dose (e.g. 35% *vs* 43% survival for NBS and GM43 cells, respectively, at 300 μg/L arsenite).

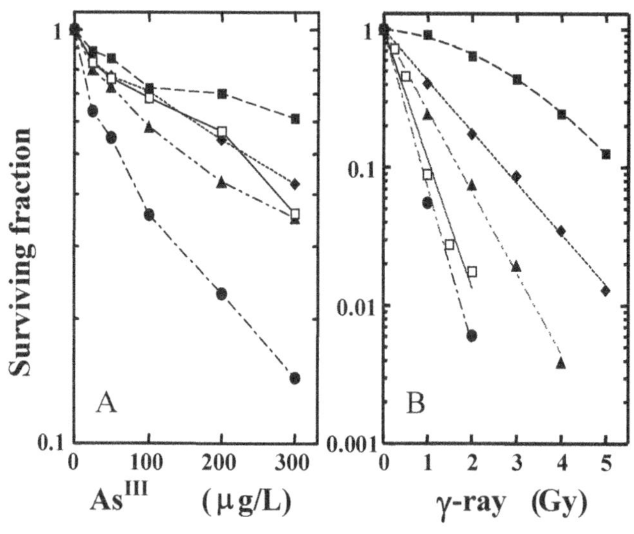


Figure 2.6 Response of double-strand break repair deficient cells to (A) and (C) 24-h exposure to sodium arsenite and (B) γ-radiation. Standard colony forming assay was used to examine the survival of the following cells (◆) GM43 (normal human fibroblast), (●) AT5BI (AT fibroblast), (▲) 780816 (NBS cells), (□) MO59J (DNA-PK deficient human glioblastoma cells, (■) MO59K (control human glioblastoma cells), (Δ) AA8 (wild type CHO cells), (◊) irs1SF (CHO cells with mutated *XRCC3* gene), (∇) V79 (wild type hamster lung fibroblasts) and (O) irs1 (V79 cells with mutated *XRCC2* gene). (*Source:* Mei et al. 2003)

Irs1 and irs1SF hamster cells carry mutations in *XRCC2* and *XRCC3*, genes involved in homology-driven double-strand break repair (Cartwright et al. 1998, Johnson et al. 1999, Liu et al. 1998). Both mutant cell lines are radiosensitive (Debenham et al. 1988, Fuller and Painter 1988), but Figure 2.6C indicates that the mutant cell lines display very similar dose responses to their respective parental cell lines, V79 and AA8, when treated with $As^{III}$. Interestingly, the V79 cells, which are derived from hamster lung, appear to be considerably more sensitive to $As^{III}$ than the ovarian AA8 cells.

## DNA Double-Strand Break Induction by Sodium Arsenite

Because of the difference in response displayed by AT cells and the double-strand break repair deficient cell lines, we examined the effectiveness of sodium arsenite to induce double-strand breaks. Of the various techniques currently available to assay double-strand breaks, we chose an immunochemical approach involving quantification of phosphorylation of histone H2AX because of its sensitivity. Within minutes of the introduction of double-strand breaks in cellular DNA, histone H2AX becomes phosphorylated at serine 139 at the carboxy terminus of the protein and forms foci (Rogakou et al. 1998), which can be readily detected by fluorescent immunostaining with antibodies to the phosphorylated form of the protein. The fluorescent intensity correlates well with the number of DNA double-strand breaks (Sedelnikova et al. 2002). Accordingly, GM34 cells were either incubated with 1000 μg/L sodium arsenite for 24 h or irradiated with 1 or 2 Gy. After cessation of treatment the cells were incubated for 30 min and then fluorescently stained for phosphorylated H2AX foci. Up to six individual cells were chosen at random from each group of cells, and the foci detected and total fluorescence intensity per cell quantified by confocal microscopy and specifically designed software. The results, which are presented in Table 2.1, indicate a linear dose response for irradiated cells with 2-Gy irradiated cells displaying an almost 10-fold increase in fluorescence over untreated cells. On the other hand, no significant increase in fluorescence was observed in the As-treated cells, despite the fact that the dose of sodium arsenite used was more toxic than irradiation with 2 Gy. This strongly suggests that sodium arsenite at doses $\leq$ 1000 μg/L is a poor inducer of DNA double-strand breaks.

Table 2.1
Fluorescence intensity of phosphorylated histone H2AX

| Cell treatment | Fluorescence intensity ± std. err. (arbitrary units) | P value (student's t-test) |
|---|---|---|
| Untreated | 681 ± 186 | |
| 1 Gy | 3162 ± 690 | 0.01 |
| 2 Gy | 6708 ± 1504 | 0.0018 |
| 1000 μg/L $As^{III}$ | 717 ± 322 | 0.92 |

*Source:* Mei et al. 2003

## Changes in p53 Protein Levels in Response to Sodium Arsenite

In a recent report, Yih and Lee (2000) showed that, unlike normal cells, the AT cell line GM3395 failed to accumulate p53 protein in response to arsenite exposure. We therefore wanted to confirm this observation with the two AT fibroblast lines we had tested for arsenite cytotoxicity. The results, shown in Figure 2.7A, indicate that there was no discernable enhancement in p53 in either AT2BE or AT5BI cells following increasing exposure to arsenite. By comparison, the repair-proficient cells, A549 and GM43, responded by increasing their cellular levels of p53. Further analysis of p53 revealed that exposure of A549 cells to $As^{III}$, unlike UV and γ-radiation, did not lead to phosphorylation of serine 15 (Figure 2.7B).

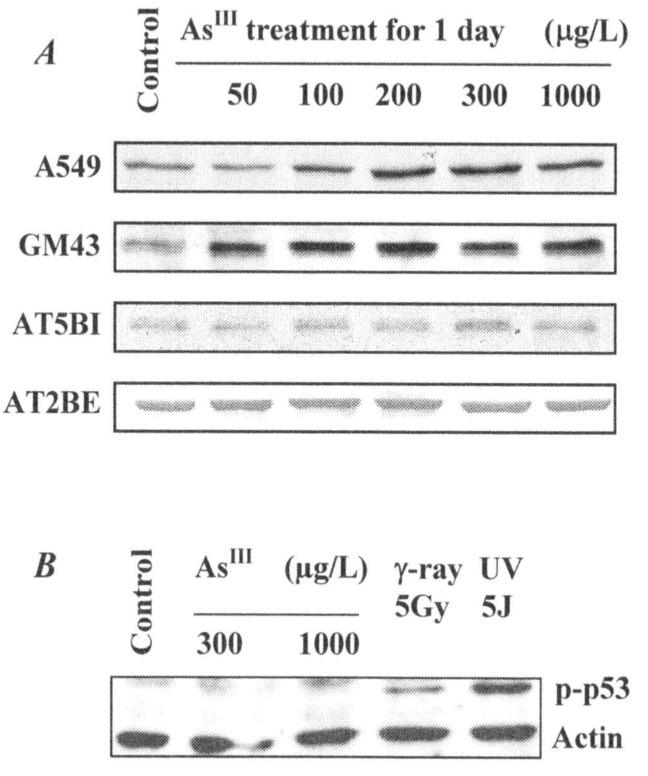


Figure 2.7 (A) Western blot analysis of p53 protein levels in AT and repair proficient cells after exposure to sodium arsenite. (B) Western blot of phosphorylation of p53 serine 15 in A549 cells in response to exposure to $As^{III}$, γ-radiation and UV radiation. The faint bands migrating slower than phosphorylated p53 (p-p53) in the lanes of control and $As^{III}$-treated cells are due to non-specific binding by the antibody. (*Source:* Mei et al. 2003)

**Influence of Sodium Arsenite on the Cell Cycle**

GM43 cells and AT5BI cells were examined for alterations to the percentage of cells in each phase of the cell cycle following 24-h exposure to sodium arsenite at 300 and 1000 μg/L. The results shown in Figure 2.8 indicated that $As^{III}$ caused a significant increase in the percentage of GM 43 cells in S-phase and a modest increase in cells in G2/M phase. By comparison, the cell cycle distribution of the AT cells appeared to be unaffected by arsenic exposure.

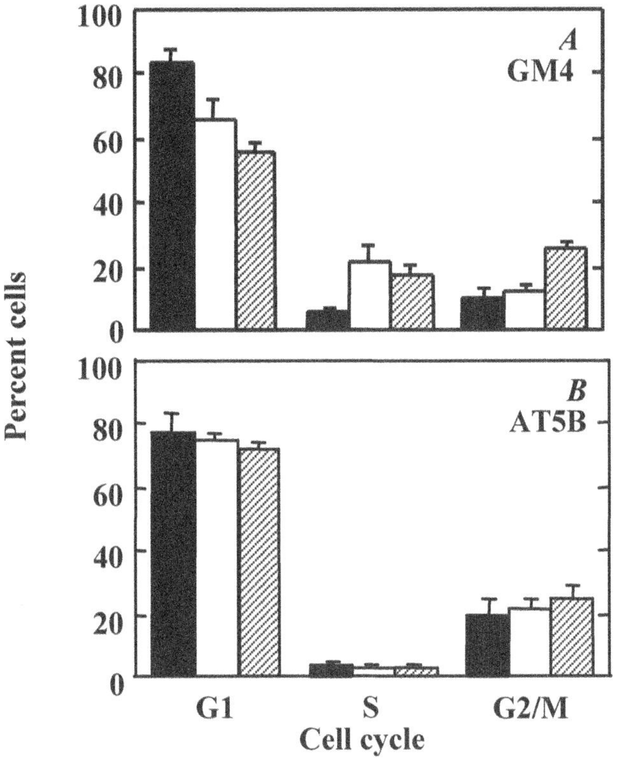


Figure 2.8. The effect of $As^{III}$ on cell cycle distribution of GM43 (A) and AT5BI (B) cells. Exponentially growing cells were incubated for 24 h with 300 μg/L (clear bars) or 1000 μg/L (hatched bars) $As^{III}$ or mock treated (solid bars) and then fixed, stained and counted as described in Materials and Methods. The percentages of cells in different phases represent the mean ± SD from three experiments. (*Source:* Mei et al. 2003)

## DISCUSSION

To date oxidative damage to DNA by arsenite has been inferred through the use of free radical scavengers, and from measurement of strand break induction before and after cleavage of the DNA by an enzyme, such as formamidopyrimidine-DNA glycosylase, that recognizes the oxidative lesion 8-oxoguanine. However, this enzyme will also cleave DNA at abasic sites, which can be introduced by other chemical and biochemical processes. Here, we have shown that arsenite-treatment of cells leads to the formation of Tg. This not only provides direct evidence for the induction of oxidative DNA damage by $As^{III}$, but also indicates that hydroxyl radicals are among the DNA damaging species generated directly or indirectly by $As^{III}$, since these are the oxidative free radicals most likely to form Tgs (Cadet et al. 1997). The elevated presence of hydroxyl radicals in arsenite-treated cells was recently demonstrated by spin-trapping and ESR (Liu et al. 2001b). Damage to DNA by OH radicals could explain the mutagenic activity of $As^{III}$ and the ability of the free radical scavenger, DMSO, to markedly reduce arsenite mutagenicity (Hei, Liu, and Waldren 1998).

Since we obtained direct evidence that arsenite generates oxidative DNA damage in human cells, an appropriate follow up question was to ask whether human cell lines, previously associated with abnormal responses to oxidative damage, would show enhanced sensitivity to $As^{III}$. Recent evidence suggests that FA cells have a reduced capacity to repair oxidative DNA damage (Lackinger 1998). BS cells are characterized by a high level of chromosomal instability and oxidative stress (Nicotera 1991). The gene defective in Bloom's syndrome (BLM) codes for a DNA helicase, which may be involved in resolution of recombination intermediates during replication or DNA repair (Karow et al. 2000). AT cells are well known for their sensitivity to ionizing radiation and other free radical inducing agents (Rotman and Shiloh 1999). We also examined three XP cell lines from different complementation groups. Although the XP proteins are primarily associated with the nucleotide excision repair pathway and therefore the repair of bulky lesions, such as UV-induced cyclobutanepyrimidine dimers, several reports have indicated XP cell sensitivity to oxidative damage, in particular cells from XP complementation groups A and G (Driggers et al. 1996, Le Page et al. 2000). However, of these particular cell types, only the AT cells displayed hypersensitivity to arsenite (Figures 2.5 B and C). The normal sensitivity of all three XP cell lines to $As^{III}$ rules out the possibility that arsenic itself generates cytotoxic bulky DNA lesions, although it has previously been shown that arsenic inhibits the nucleotide excision repair pathway (Okui and Fujiwara 1986, Hartwig et al. 1997).

The pattern of sensitivity of the AT cell lines to arsenite mimicked the response to radiation. This included the SV40-transformed AT cell lines which showed intermediate sensitivity to both agents. The restoration of the AT cells to a normal response to $As^{III}$ by complementation with ATM cDNA expression confirmed the importance of the role of ATM in cellular defense against this agent. The ATM protein is known to be intimately involved in response to genotoxic insult and cell cycle regulation (Rotman and Shiloh 1999). It is a protein kinase and, like DNA-PK, belongs to a family of proteins that possess a region similar to the catalytic domain of phosphatidylinositol 3-kinase. One of the substrates for ATM phosphorylation after cellular irradiation is NBS1 (Lim et al. 2000, Gatei et al. 2000, Zhao et al. 2000). Surprisingly, we observed that NBS1 cells did not display a marked sensitivity to $As^{III}$, although they displayed the expected sensitivity to radiation (Figure 2.6). This implies that ATM does not act through NBS1 in response to arsenite and also that the homology-driven double-strand break repair pathway involving the NBS1/Mre11/Rad50 complex is not required for protection against $As^{III}$. Furthermore, since the cytotoxicity of the DNA-PK, XRCC2, and XRCC3 deficient cell lines (MO59J, irs1, and irs1SF cell lines, respectively) after exposure to $As^{III}$ fell in the normal range, it appears that double-strand breaks may not contribute significantly to arsenite cytotoxicity, at least for arsenite concentrations $\leq 300\ \mu g/L$. This was further substantiated when we observed that $As^{III}$ did not generate significant levels of double-strand breaks.

The ATM protein phosphorylates p53 and MDM2 in response to DNA damage by agents such as ionizing radiation (Banin et al. 1998, Canman et al. 1998), which in turn leads to the accumulation of p53 in the cell (Jiminez et al. 1999). Like other laboratories, we observed an accumulation of p53 in normal cells (Yih and Lee 2000, Vogt and Rossman 2001, Menendez et al. 2001) but not in the AT cells (Yih and Lee 2000, Menendez et al. 2001) following treatment with arsenite (Figure 2.7A). Thus it would appear that ATM acts on p53 in response to arsenite exposure. However, unlike the response seen after $\gamma$ irradiation, serine 15 of p53 was not phosphorylated (Figure 2.7B). Phosphorylation of this residue has been associated with DNA double-strand break induction. For example, Nakagawa et al. (1999) reported phosphorylation of serine 15 following microinjection of a restriction endonuclease. Thus, the lack of p53 phosphorylation at serine 15 provides further evidence that As cytotoxicity is not the result of double-strand break induction.

Cumulatively, our observations with DNA repair deficient cell lines, as well as direct measurement of double-strand breaks and other cellular responses that frequently accompany double-strand breaks, call into doubt whether DNA damage is the primary cause of death when cells are exposed to arsenic. As discussed, bulky DNA adducts requiring the nucleotide excision repair pathway can be ruled out because of the lack of abnormal sensitivity shown by the XP cells. Similarly, the cell survival curves for EM9, NBS, MO59J, irs1 and irs1SF cells argue against the importance of oxidative base damage, single-strand breaks or double-strand breaks. Fanconi anemia cells and XRCC2 and XRCC3 mutated cells are also known to be hypersensitive to cross-linking agents that cause DNA interstrand crosslinks, such as mitomycin C (Fujiwara 1982, Jones et al. 1987, Caldecott and Jeggo 1991). Their normal response to arsenite suggests that DNA interstrand crosslinks do not contribute to arsenic toxicity. DNA-protein crosslinks are the only remaining major class of DNA lesions that cannot be ruled out by our experiments, and several groups have shown that these lesions can be generated by exposure to arsenite (Dong and Luo 1993, Gebel et al. 1998, Wang et al. 2001). Although this will be the subject of further investigation, preliminary examination of an ERCC1 mutant of AA8 (UV20), which may be sensitive to DNA-protein cross-linking agents such as irradiation under hypoxia (Murray and Rosenberg 1996), indicated no elevated sensitivity to arsenite (data not shown).

As an alternative to DNA damage, arsenic toxicity may be mediated by altering the normal course of the cell cycle. The response of normal fibroblasts to accumulate in S phase (Figure 2.8), at least transiently, after treatment with arsenite has been noted before (Yih and Lee 2000). It is probably attributable to a block in DNA synthesis caused by $As^{III}$ (Ochi, Nakajima, and Fukumori 1998). That the AT cells fail to show a similar cell cycle response was not surprising, since one of the hallmarks of AT cells is their failure to undergo inhibition of DNA synthesis following insult such as cellular irradiation (Painter 1981). The results we have obtained with sodium arsenite are reminiscent of recent observations regarding the cellular response to t-butyl hydroperoxide (Shackelford et al. 2001), including the failure of AT fibroblasts to induce p53 when exposed to this agent.

Although we have used cell survival as our major endpoint, this study has implications for arsenic carcinogenesis because AT is a cancer prone disorder. The question of the cancer predisposition of AT heterozygotes, who constitute approximately 1% of the general population, remains controversial (Su and Swift 2000; Gatti, Tward, and Concannon 1999). However, if it is confirmed that AT heterozygotes are over-represented in cancer patient populations, it will be important to identify potential carcinogens to which AT heterozygotes are more susceptible. The results of this and other studies (Yih and Lee 2000, Menendez et al. 2001) would suggest that arsenic, because of its prevalence, could belong to this group of agents.

# CHAPTER 3
# ADDITIVE EFFECTS OF SODIUM ARSENITE AND γ-RADIATION

## INTRODUCTION

Exposure to multiple agents may be critical in defining the toxic and carcinogenic potentials of a particular agent. Synergistic and additive toxic effects of arsenicals in cellular systems still need to be fully defined (Abernathy et al. 1999). Arsenic has been shown to enhance cytotoxicity, clastogenecity and mutagenicity in combination with other DNA damaging agents, including UV radiation, alkylating agents and benzo[*a*]pyrene. Post-treatment with As(III) synergistically increased the cytotoxity of UV light in Chinese hamster ovary (CHO) cells (Lee et al. 1991). When UV-irradiated plasmid was allowed to replicate in human fibroblasts treated with $As^{III}$, the yields of mutations were significantly greater than the yield expected if the effects of each treatment were simply additive (Wiencke et al. 1997). Studies have also indicated a synergistic interaction between arsenic exposure and cigarette smoking in the induction of lung cancer (Tsuda et al. 1995). But, little is known of the interaction between arsenic and γ-radiation. It was reported that sodium arsenite synergistically enhanced the frequency of X-ray-induced chromatid aberrations in human lymphocytes from two donors (Jha et al. 1992). A more recent study indicated a synergistic interaction between arsenic trioxide fractionated radiotherapy in the treatment of a murine fibrosarcoma, although this was most likely due to a reduction in tumor blood flow (Lew et al. 2002). Vogt and Rossman (2001) found that 14-day exposure to a low concentration of arsenite significantly suppressed p21 induction by ionizing radiation.

Low concentration arsenic exposure has become of more concern than high concentration exposure mainly because of its ubiquitous existence. We have, therefore, investigated the effects of low concentrations of $As^{III}$ in combination with γ-radiation in human cells by examining the repair of radiation-induced DNA base damage and cell survival. We again made use of the panel of human cell lines described in the previous chapter to determine whether differing DNA repair capacities further influenced arsenic-radiation interaction.

## RESULTS

### Effect of $As^{III}$ on the Removal of Thymine Glycol in A549 Cells

Thymine glycol (Tg) is a DNA lesion produced by ionizing radiation and other oxidative agents. We employed the capillary electrophoresis-based immunoassay for Tg (see previous chapter) to determine if $As^{III}$ influences the cellular removal of the damaged base from the DNA of irradiated cells. A549 cells were preincubated for 24 h in media containing 50 μg/L sodium arsenite prior to irradiation with 2 Gy and then allowed to carry out DNA repair for up to 8 h still in the presence of $As^{III}$. (The dose of $As^{III}$ was chosen on the basis of our previous data showing that the level of DNA damage produced by 50 μg/L $As^{III}$ is substantially lower than that produced by 2-Gy irradiation.) The control cells were simply irradiated without exposure to $As^{III}$, either before or after irradiation. The results are shown in Figure 3.1. In agreement with our previously published data (Le et al. 1998), A549 cells removed over 80% of the radiation-induced Tg within 4 h. However, the presence of $As^{III}$ clearly slows the rate of removal Tg removal.

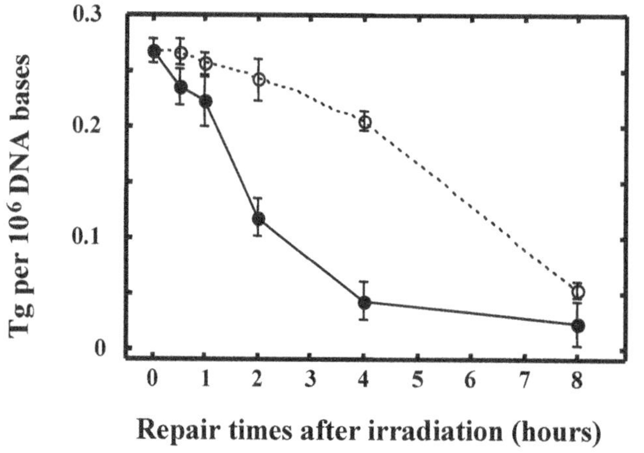


Figure 3.1 Effects of $As^{III}$ on the removal of thymine glycol (Tg) in A549 cells. The A549 cells were exposed to 2 Gy of γ-irradiation with (○) or without (●) pre-incubation with $As^{III}$ of 50 μg/L for 1 day then incubated at 37°C for different repair times.

### Cytotoxicity of Human Cell Lines to $As^{III}$ and γ-Radiation as Single Agents

Cytotoxicity resulting from exposure to arsenic and γ-radiation was examined in a panel of eight human cell lines. We compared the response of three DNA repair-proficient cell lines, A549 (human lung carcinoma), GM38 and GM43 (normal human fibroblasts) to five repair-deficient cell lines CRL-1223 (fibroblasts derived from a patient with Xeroderma pigmentosum complementation group A), GM434 (XP-D fibroblasts), GM3021 (XP-G fibroblasts), FA1196 (Fanconi anemia fibroblasts) and AG06040 (Bloom's syndrome fibroblasts). Representative survival curves are illustrated in Figure 3.2. The response to sodium arsenite was examined in two ways, either after 24-h exposure to doses up to 300 μg/L (Figure 3.2A) or after incubation with 300 μg/L for times up to 6 days (Figure 3.2B). Figure 3.2a indicates that, at the relatively low doses used for this study, the survival curves are mostly exponential (i.e. a straight line in a semi-logarithmic plot) with little evidence for a pronounced initial shoulder to the curves. Although the cells exhibited a fairly broad range of sensitivities to arsenite, the repair-deficient cells showed no marked hypersensitivity compared to the survival range defined by the three repair-proficient cells. The surviving fractions (SF%) after 300 μg/L $As^{III}$ treatment for 24 hours were 42% for CRL-1223, 52% for GM434, 50% for GM3021, 67% for FA1196 and 39% for AG06040 compared with 61% for A549, 79% for GM38 and 42% GM43 cells (Figure 3.2A). Prolonged incubation did not drastically alter the rank order of cell sensitivities except for the pronounced relative increase in sensitivity displayed by one of the repair-proficient cell lines, GM38 (Figure 3.2B). No relationship between sensitivity to $As^{III}$ and γ-radiation was observed (Figures 3.2A and 3.2C). The repair-deficient cell lines appeared to be only marginally more sensitive to γ-radiation compared to GM38 and GM43 cells. A549 cells are known to be radioresistant (Weinfeld et al. 1997).

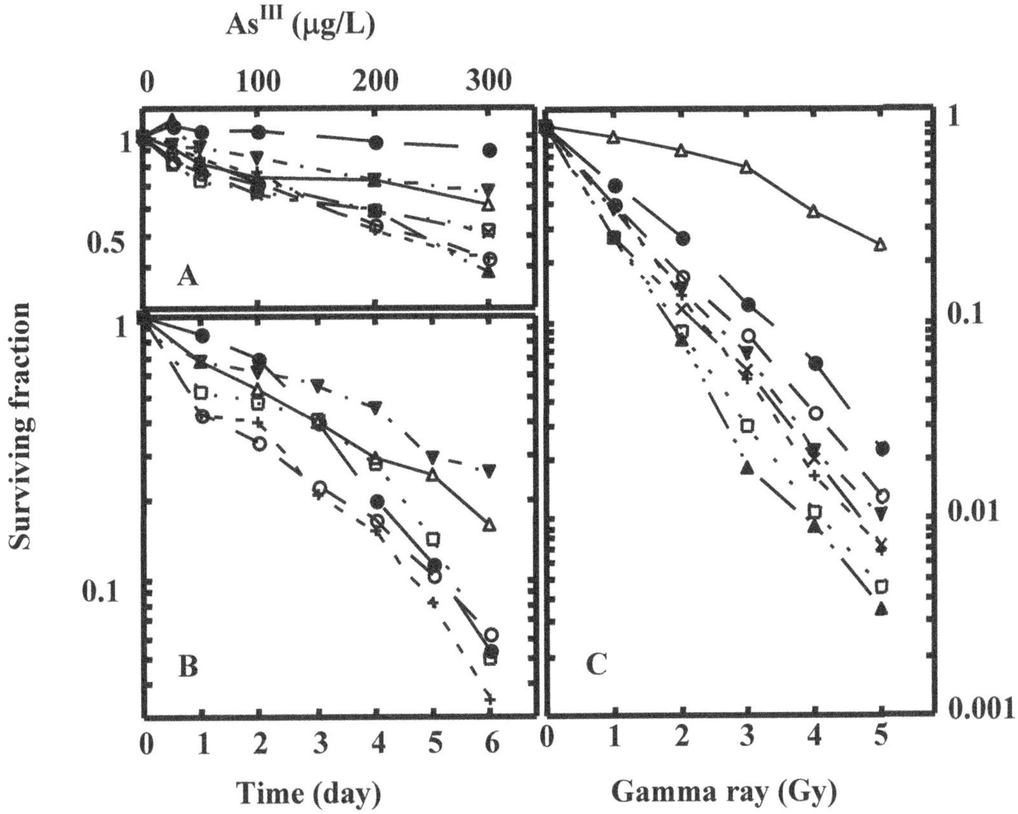


Figure 3.2 Cell survival of various cell strains after treatments. The cells were plated and allowed to attach for a period of 16~24 hours, and then treated with different doses of γ-radiation or $As^{III}$. The cells were grown in monolayer culture at 37°C and 5% $CO_2$ for 2 weeks. The data are taken from 4~8 survival curves. **(A)** cell survival after $As^{III}$ treatment for 1 day; **(B)** cell survival after 300 μg/L of As(III) treatment for different times; **(C)** cell survival after γ-irradiation. (Δ), A549; (●), GM38; (○), GM43; (+), CRL-1223; (□), GM434; (X), GM3021; (▼), FA1196; (▲), AG06040.

### Cytotoxicity of the Combination of $As^{III}$ and γ-Radiation

Figure 3.3 shows the effects of $As^{III}$ on 2-Gy γ-irradiated cell lines. The cells were exposed to 2 Gy (the dose usually delivered per fraction in radiotherapy) and then incubated with increasing concentrations of $As^{III}$ for 24 hours. The results of the converse experiment, in which the cells were exposed to increasing doses of radiation from 1 to 5 Gy and then incubated with a single dose (300 μg/L) of sodium arsenite for 24 hours, are shown in Figure 3.4. In both circumstances the curves for the combined treatments appear to run more or less parallel to the curves of the exposure to the single agents. Furthermore the results appear to match closely the additive effects expected from Figure 3.2A and 3.2C. For example, the decrease in surviving fraction of GM43 cells at each arsenic dose due to prior irradiation with 2 Gy is of the order of 80 % (Figure 3.3), which is in agreement with the 20% survival shown for GM43 after 2-Gy irradiation alone (Figure 3.2C). Similarly the decrease in GM43 survival at each radiation dose is 60% (Figure 3.3), in line with the 60% toxicity inflicted by sodium arsenite alone (Figure 3.2A). A statistical analysis of the data in Figure 3.3 is shown in Table 3.1.

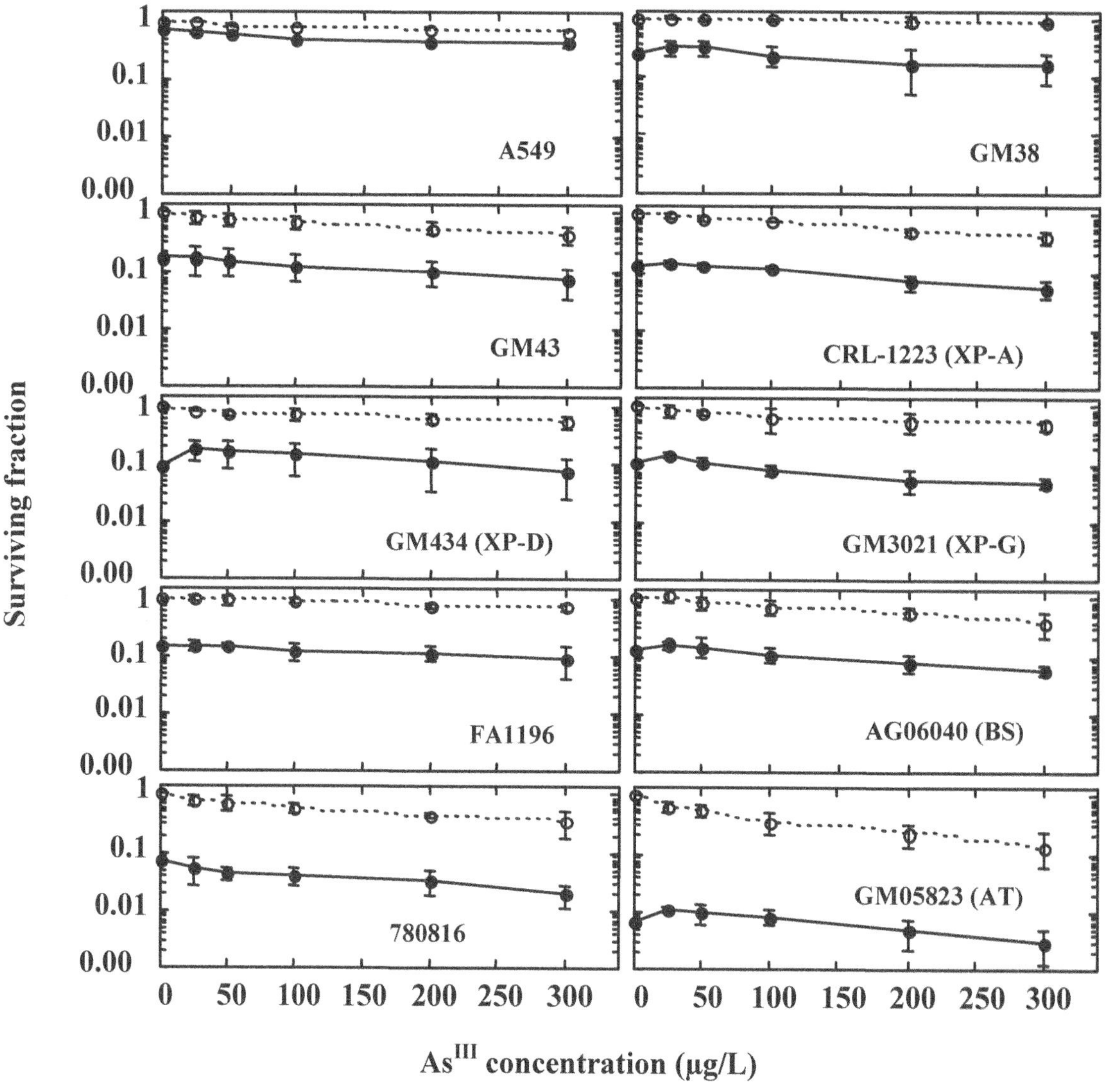


Figure 3.3 Effects of As$^{III}$ on 2 Gy γ-irradiated cells of different cell lines. The cells after attachment were treated with different concentrations As$^{III}$ for 1 day with (●) or without (○) pre-exposure to 2 Gy of γ-radiation. Data are presented as means ±SD from 3-6 experiments.

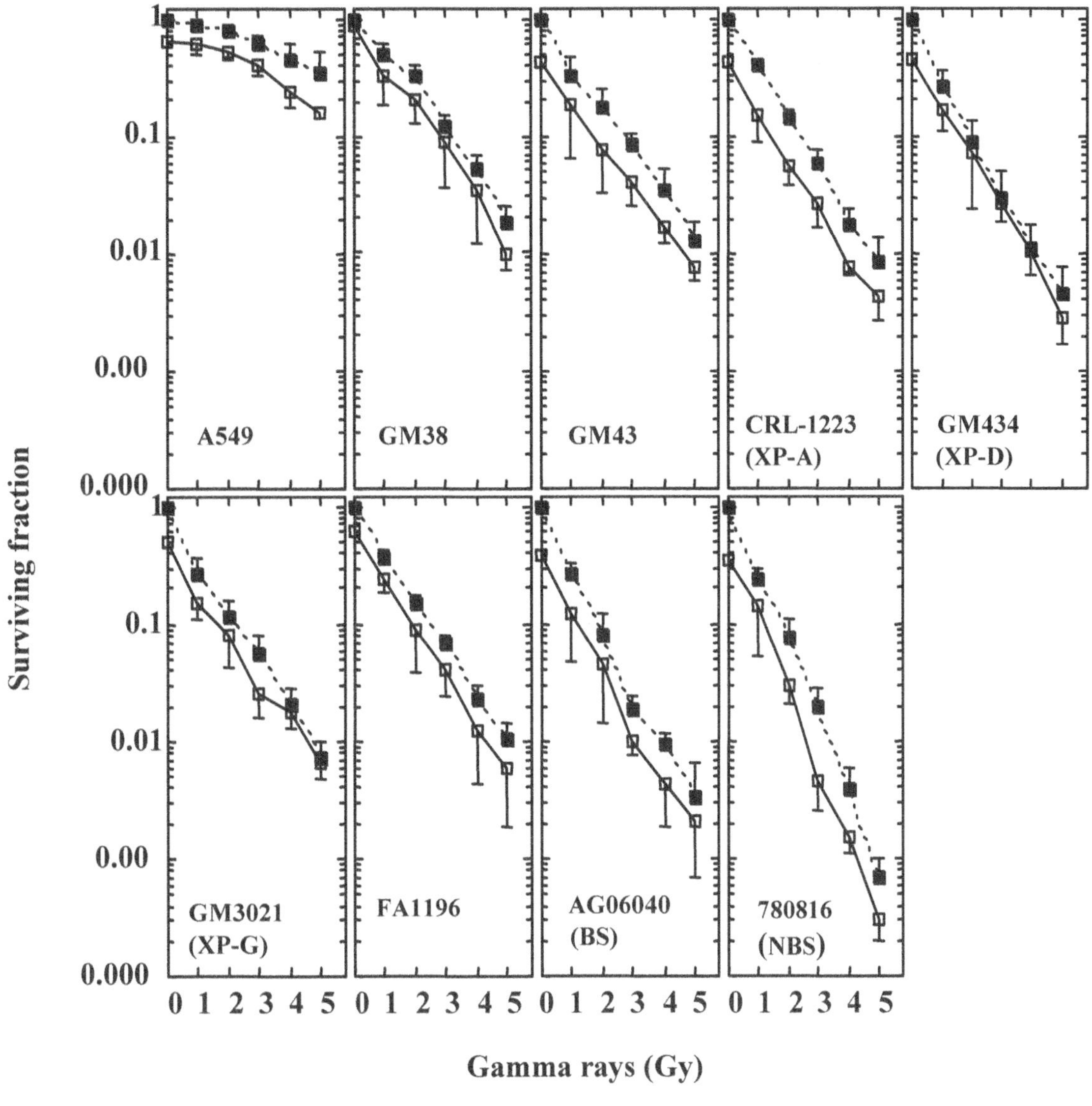


Figure 3.4 Effect of 300 μg/L $As^{III}$ on γ-irradiated cells. The cells were exposed to 1-5 Gy of γ-radiation with (□) or without (■) post-treatment with 300 μg/L of $As^{III}$ for 1 day. Data are presented as means ± SD from 3-6 experiments.

Table 3.1
Parameters determined for the interaction of 2-Gy irradiation with $As^{III}$ exposure

| Cell line | 2 Gy exposure | Slope of survival curve* | P value for difference between slopes† | Calculated average difference in surviving fraction |
|---|---|---|---|---|
| A549 | - | -0.00147 | | |
| | + | -0.00166 | 0.683 | 0.31 |
| GM38 | - | -0.00047 | | |
| | + | -0.00049 | 0.022 | 0.62 |
| GM43 | - | -0.00266 | | |
| | + | -0.00292 | 0.357 | 0.80 |
| CRL-1223 | - | -0.00294 | | |
| | + | -0.00359 | 0.119 | 0.84 |
| GM434 | - | -0.00189 | | |
| | + | -0.00194 | 0.967 | 0.83 |
| GM3021 | - | -0.00208 | | |
| | + | -0.00350 | 0.100 | 0.85 |
| FA1196 | - | -0.00130 | | |
| | + | -0.00171 | 0.037 | 0.85 |
| AG06040 | - | -0.00328 | | |
| | + | -0.00326 | 0.973 | 0.84 |

Note:
*: For the regression analysis see Sokal and Rohlf (1969)
†: For the statistical analysis software: SAS Software Release 6.12, SAS Institute Inc., Cary, NC, USA

For the arms of each experiment a linear regression equation was fitted to the log survival as a function of the dose and a test for equality of slopes was performed comparing the control arm to each of the radiation response arms. We found that there was no statistically significant difference between the slopes ($p < 0.05$) except for GM38 and FA1196. In these cases, although the difference was statistically significant, the absolute amount of the difference was small. A similar analysis of the data in Figure 3.4 (data not shown) indicated that, except for A549, there was no statistically significant difference between the slopes. For A549 cells the radiation response was best described by a linear quadratic equation and a comparison of the quadratic terms indicated a small, but statistically significant difference ($p = 0.033$).

We also examined the influence of arsenic treatment time on radiation survival response (Figure 3.5). GM38 and CRL-1223 cells were irradiated with 1-5 Gy $\gamma$-radiation and then incubated with 300 μg/L $As^{III}$ for 4 h, 1 and 3 days. With increasing treatment times, the $As^{III}$ reduced cell survival beyond that of the radiation alone, but in an additive manner consistent with the $As^{III}$-induced cytotoxicty displayed in Figure 3.2B.

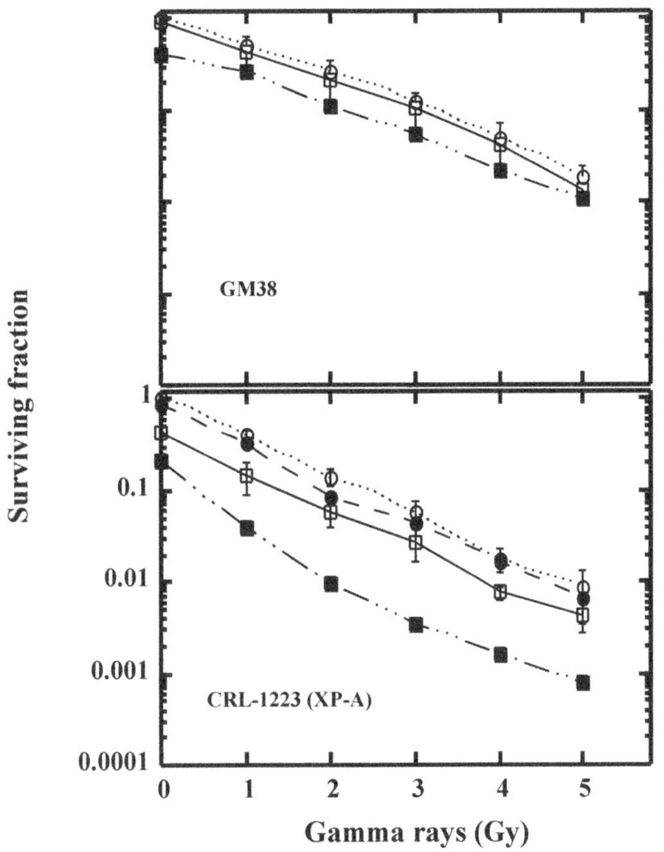


Figure 3.5 Effects of 300 μg/L As$^{III}$ on γ-irradiated GM38 and CRL-1223 (XP-A) cells. The cells were exposed to 1-5 Gy of γ-radiation without (○) or with post-treatment with 300 μg/L of As$^{III}$ for 4 h (●), 1 day (□) and 3 days (■). Data points are presented as the means ±SD from 4-6 experiments.

## Comparison Between Pre- and Post-Treatment with As$^{III}$ in γ-Irradiated Cells

To assess the effects of the timing of the arsenite treatment with respect to the irradiation, three cell lines (A549, GM3021 and 780816) were irradiated with 1-5 Gy before or after 24-h treatment with 300 μg/L sodium arsenite. Figure 3.6 shows the results. The cell killing effects of γ-radiation were enhanced by As$^{III}$ in an additive manner. No significant differences between pre- and post-treatment with As$^{III}$ were observed. We have also complemented this study by using different concentrations of As$^{III}$ with γ-irradiated A549 cells. The A549 cells were irradiated with 2 Gy before or after incubation with As$^{III}$ treatment for 4 or 24 hours using concentrations ranging from 25 to 300 μg/L. No significant differences between pre- and post-treatment with As$^{III}$ were observed (data not shown).

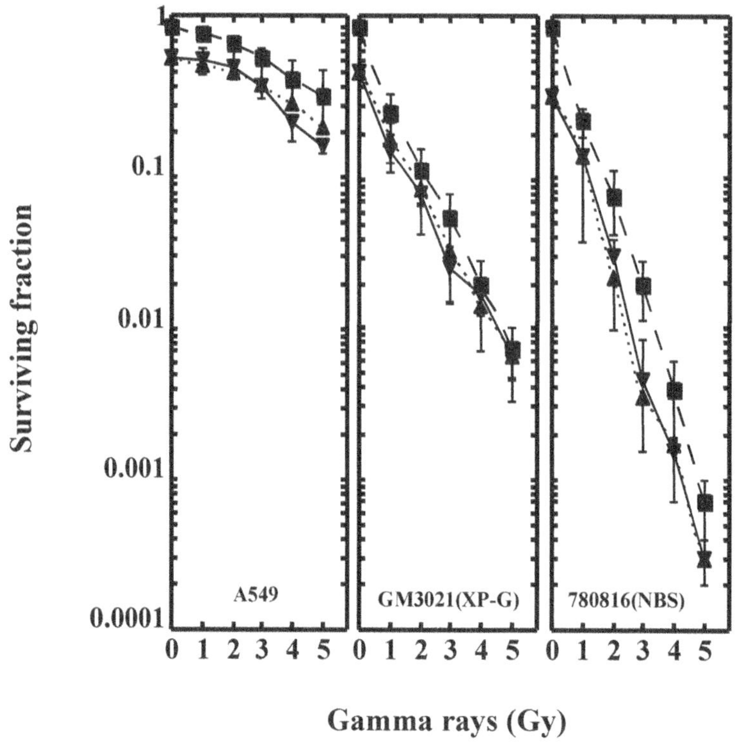


Figure 3.6 Comparison of cell survival of pre- vs. post- treatment with $As^{III}$ for 1 day with γ-irradiated cells. (■) γ-radiation only; (▼) γ-radiation with post-treatment of 300 μg/L $As^{III}$; (▲) γ-radiation with pre-treatment of 300 μg/L $As^{III}$.

**Effects of Low Concentrations of $As^{III}$**

Figure 3.7 shows the effects of different times and concentrations of $As^{III}$ as low as 25 μg/L on γ-irradiated CRL-1223 cells. After exposure to 1-5 Gy γ-radiation, the cells were incubated with 25 μg/L $As^{III}$ for 12 days or 50 μg/L $As^{III}$ for 6 days compared to 300 μg/L $As^{III}$ for 1 day. A significant additive effect could not be detected with the lower $As^{III}$ concentrations.

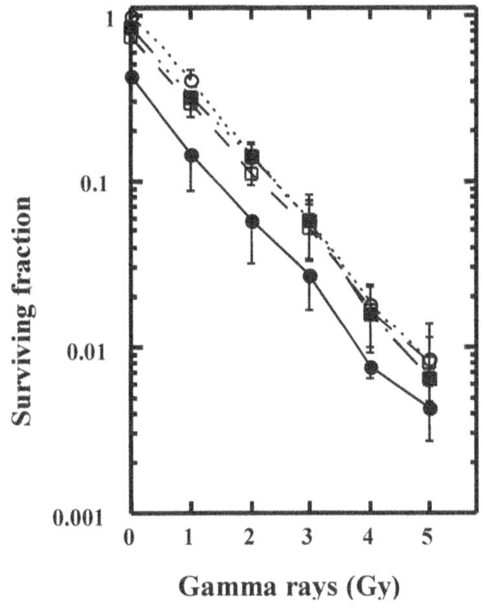


Figure 3.7 Effects of different concentrations and times of $As^{III}$ in γ-irradiated CRL-1223 (XP-A) cells. After exposure to 1-5 Gy of γ-radiation, CRL-1223 cells were incubated with or without $As^{III}$ of different concentrations for different times. (○) without $As^{III}$; (●) 300 µg/L of As for 1 days; (□) 50 µg/L for 6 days; (■) 25 µg/L for 12 days. Data points are presented as means ±SD from four experiments.

## DISCUSSION

In the previous chapter we showed that arsenic can cause oxidative DNA damage to cellular DNA. An alternative mechanism by which arsenic could enhance mutation frequency and neoplastic transformation is by interfering with the repair of DNA lesions, including lesions that might result from endogenous cellular oxidation. Arsenic has been shown to enhance mutagenicity and genotoxicity in combination with other DNA damaging agents such as alkylating agents and UV light by inhibiting DNA repair (Abernathy et al. 1999, Hartwig et al. 1997, Hartwig 1998, Lynn et al. 1997). Here we have provided evidence that sodium arsenite, at the relatively low concentration of 50 µg/L, slows the removal of thymine glycol from irradiated cells. It remains to be determined whether this is mediated by direct interference of hNTH, the DNA glycosylase primarily responsible for removal of thymine glycol in human cells, or by some other mechanism.

Given the inhibitory effect of $As^{III}$ to the base repair machinery observed at the molecular level, it was of interest to see if this would translate into a cytotoxic interaction at the cellular level. Our comprehensive analysis of the co-treatment of cells with radiation and arsenite revealed that under almost all circumstances examined cytotoxicity was purely additive. There was no indication of a synergistic interaction even in the repair-deficient mutant cell lines. This strongly implies that ionizing radiation and arsenic kill cells by distinct mechanisms. Furthermore, it implies that slowing down the repair of DNA base damage, such as thymine glycols, does not enhance radiation-induced cell killing, reinforcing the idea that double-strand breaks and other complex lesions, rather than simple base lesions, are the most likely cytotoxic lesions induced by radiation (Ward 1994). Importantly, however, it is clear from the previous chapter that the ATM protein plays a key role in protecting cells from both radiation and arsenic. In the case of radiation, structural changes to chromatin, possibly as a result of DNA double-strand breaks, may serve as the signal for ATM activation (Bakkenist and Kastan 2003). How arsenic activates ATM remains to be determined, but it is possible that it may not even involve DNA damage.

# CHAPTER 4
# ASSAY FOR DNA DAMAGE USING CAPILLARY ELECTROPHORESIS LASER-INDUCED FLUORESCENCE

## INTRODUCTION

The studies described in Chapters 2 and 3 required sensitive techniques for detection of trace amounts of DNA damage. This chapter describes our efforts toward the development of highly sensitive immunoassay for DNA damage.

Structural damage to DNA is generally considered to be the initial step in the complex multistage model of cancer development (Friedberg, Walker, and Siede 1995). Human DNA is exposed to a variety of endogenous and environmental agents that may induce a wide range of damage. Many of these DNA damaging agents, including polycyclic aromatic hydrocarbons (PAHs), produced by incomplete combustion of organic materials, induce DNA lesions or form adducts in the DNA structure. One of this class of compounds, benzo[*a*]pyrene, has been extensively studied due to its strong ability to react with DNA and cause mutation. Benzo[*a*]pyrene is metabolized *in vivo* by cytochrome P450 and epoxide hydrolase to form benzo[*a*]pyrene-7,8-diol 9,10-epoxide (BPDE), which is generally considered to be the reactive carcinogenic species. The reactive epoxide moiety of BPDE binds to DNA, primarily to the $N^2$ position of deoxyguanosine, to form a bulky adduct (BPDE-$N^2$-dG) (Szeliga and Dipple 1998). This adduct causes changes in the conformation of the DNA helix surrounding the damaged site. This distortion can disrupt biological processes including DNA replication and transcription (Hess et al. 1997, Choi et al. 1994, Thrall et al. 1992), thus requiring DNA repair. Mutations by BPDE have been shown to affect genes critical to the development of cancer, including the tumor suppressor p53 (Denissenko et al. 1996). Consequently, the detection of DNA damage caused by compounds such as benzo[*a*]pyrene is an important issue regarding human exposures to these environmental carcinogens.

An array of techniques has been developed to detect BPDE-DNA damage (Weston 1993, Cadet and Weinfeld 1993, Pfeifer 1996), including immunochemical methods (Pfeifer 1996, Poirier et al. 1980, Santella 1999, Baan et al. 1988), $^{32}$P-postlabeling (Pfeifer 1996, Beach and Gupta 1992, Zeisig and Moller 1995), single cell gel electrophoresis (comet) assays (Pfeifer 1996, Hanelt et al. 1997), PCR-based assays (Cadet and Weinfeld 1993, Pfeifer 1996), fluorescence methods (Pfeifer 1996; Pavanello et al. 1999; Li, Hurtubise, and Weston 1999), chromatography and capillary electrophoresis, and their coupling with mass spectrometry (Chiarelli and Lay 1992; Giese 1997; Barry, Norwood, and Vouros 1996; Andrews, Vouros, and Harsch 1999; Nackerdien and Atha 1996). One of the commonly used techniques is the enzyme-linked immunosorbent assay (ELISA) using antibodies specific for the type of damage being investigated. Several monoclonal antibodies that recognize BPDE-DNA damage have been developed and used for ELISA (Poirier et al. 1980, Baan et al. 1988, Santella et al. 1984). The ELISA method has been used to determine BPDE adduct levels in samples from human subjects exposed to high levels of benzo[*a*]pyrene (Santella 1999). While very useful, this technique requires relatively large amounts of DNA (200 μg or more) and the detection limit is not sufficient for monitoring BPDE adducts at the levels found in human samples from the general population (Santella et al. 1995, Kang et al. 1995, Dickey et al. 1997).

To address some of these limitations, we have developed an assay that combines immunological recognition of damaged DNA, capillary electrophoresis separation, and laser-induced fluorescence detection. A primary ($1^o$) mouse monoclonal antibody specific for the DNA lesion was used to bind to the DNA lesion. A secondary ($2^o$) anti-mouse IgG antibody that was labeled with a fluorescent dye, tetramethylrhodamine (TMR), was used to bind with the primary antibody. The resulting complex of $2^o$ antibody + $1^o$ antibody + damaged DNA was separated using free-zone capillary electrophoresis and detected with laser-induced fluorescence.

The use of antibodies to specific DNA damage provides selectivity needed for recognizing minute amounts of the DNA damage in the presence of large excess of normal DNA. Antibodies to various DNA damage types are available and can be used for DNA damage analysis. Antibodies as a class of molecules possess a high level of heterogeneity, which is key to their role in the immune system. Although the basic structure of an antibody isotype such as immunoglobulin G (IgG) is very similar, each antibody has a unique amino acid sequence in its antigen-binding site (Abbas, Lichtman, and Pober 1994). Differences in this region may include the number of charged amino acid side chains as well as variations in the surface hydrophobicity of the antibody. These characteristics may affect both the electrophoretic mobility and the interaction of the antibody with the capillary wall during free-zone capillary electrophoresis. To study antibody-DNA damage interactions and to assist the development of assays for DNA damage, we have designed a synthetic, fluorescently-labeled DNA damage standard using BPDE adducts as the model damage type.

It was necessary to design a standard DNA molecule that contains a known amount of a specific damage type, and that can be fluorescently-labeled for high-sensitivity detection by laser-induced fluorescence (LIF). We designed and generated a 90-base pair double-stranded oligonucleotide containing a BPDE-$N^2$ deoxyguanosine (dG) adduct in the middle of one strand, with a tetramethylrhodamine (TMR) fluorophore attached to the 5'-end of the same strand. This characteristic allowed for studies using either single- or double-stranded DNA to be performed using the CE/LIF system. The probe's design also allows substitution of different damage types in the oligonucleotide with relative ease. Thus, a variety of DNA lesions can be studied using the same approach.

The synthetic BPDE-DNA adduct was used as a standard in a competitive assay to determine the level of BPDE-DNA adducts in a human lung carcinoma cell line exposed to BPDE. A fluorescently labeled BPDE-DNA adduct standard and a BPDE-specific antibody are added to a sample containing unknown amount of unlabeled BPDE-DNA adduct. The unlabeled BPDE-DNA adduct and the labeled BPDE-DNA adduct compete to form complexes with the antibody. CE separation of the bound and unbound adducts allows determination of the bound concentration, which in turn is related to the amount of BPDE-DNA adduct in the sample. In contrast to other methods of performing immunoassays, CE/LIF allows rapid analysis, excellent mass sensitivity and potential for automation. The popularity of this technique in immunoassays is well reflected in numerous reports (Schultz, Huang, and Kennedy 1995; Lausch et al. 1995; Schmalzing et al. 1995; Chen and Sternberg 1994; Evangelista and Chen 1994).

## MATERIALS AND METHODS

### Reagents

Unmodified oligonucleotides were synthesized by the Department of Biochemistry DNA synthesis laboratory, University of Alberta, or by Integrated DNA Technologies (Coralville, Iowa). All oligonucleotides were purified by sequencing polyacrylamide gel electrophoresis prior to use. Purity of the oligonucleotides was confirmed by $^{32}$P-radiolabeling and gel electrophoresis. Tetramethylrhodamine (TMR) - labeled oligonucleotide was synthesized by University Core DNA Services, (University of Calgary, Calgary, Alta.). (±)-r-7,t-8-dihydroxy-t-9,10-epoxy-7,8,9,10-tetrahydrobenzo[*a*]pyrene (anti) [(±)-*anti*-BPDE] was supplied by the National Cancer Institute Chemical Carcinogen Reference Standard Repository (Midwest Research Institute, Kansas City, Miss.). Premixed polyacrylamide / bisacrylamide (19:1) solution was purchased from BioRad Laboratories (Cambridge, Mass.). Enzymes were supplied by Amersham Pharmacia Biotech (Piscataway, N.J.). Monoclonal antibodies 5D11 and 8E11 were purchased from BD PharMingen (San Diego, Calif.). Cell supernatant containing monoclonal antibody E5 (Baan et al. 1988) was kindly provided by Dr. William Watson, Shell International Chemicals BV, Shell Research and Technology Center, Amsterdam, Netherlands, and was prepared as described by Booth et al. (1994). Polyclonal mouse IgG antibody was purchased from Calbiochem (La Jolla, Calif.). Solvents and other biochemicals were supplied by Sigma Chemical (St. Louis, Miss.), Fisher Scientific (Pittsburgh, Pa.), or VWR Canlab (Mississauga, Ont.).

### Design of Probe

In order to imitate DNA damage as it occurs naturally in cellular DNA, we designed a 90-base pair double-stranded oligonucleotide. The desired characteristics of this oligonucleotide were that it be fluorescently-labeled, contain a known amount of damage, and be long enough to be recognized by a variety of antibodies and other DNA-binding proteins. The oligonucleotide consists of six overlapping, complementary oligonucleotides of varying lengths that were annealed and ligated to form a complete double-stranded 90-mer (Figure 4.1). The oligonucleotide sequences used in the current study were: oligonucleotide 1: 5'-TMR-labeled-CCTTAAGCTTCCTCAACCACTTACCATACTCGAGATT-3'; oligonucleotide 2: 5'-GAGTAT-GGTAAGTGGTTGAGGAAGCTTAAGG-3'; oligonucleotide 5: 5'-GTCATATGCCGCCTCTGA-CCTTCCTAGAATTCCATCC-3'; oligonucleotide 6: 5'-GGATGGAATTCTAGGAAGGTCAG-AGGCGG-3'. The sequence of oligonucleotide 3 and its complementary strand (oligonucleotide 4) may be changed to create a variety of desired damage types, typically with a single damaged nucleotide in the middle of oligonucleotide 3. In the current study the sequences used were: oligonucleotide 3: 5'-CCCATTATGCATAACC-3'; oligonucleotide 4: 5'-CATATGACGGTTATGCATAATGGG-AATCTC-3'. The fluorescent label (oligonucleotide 1) and damaged nucleotide (oligonucleotide 3) are on the same strand to allow both double- and single-stranded DNA studies.

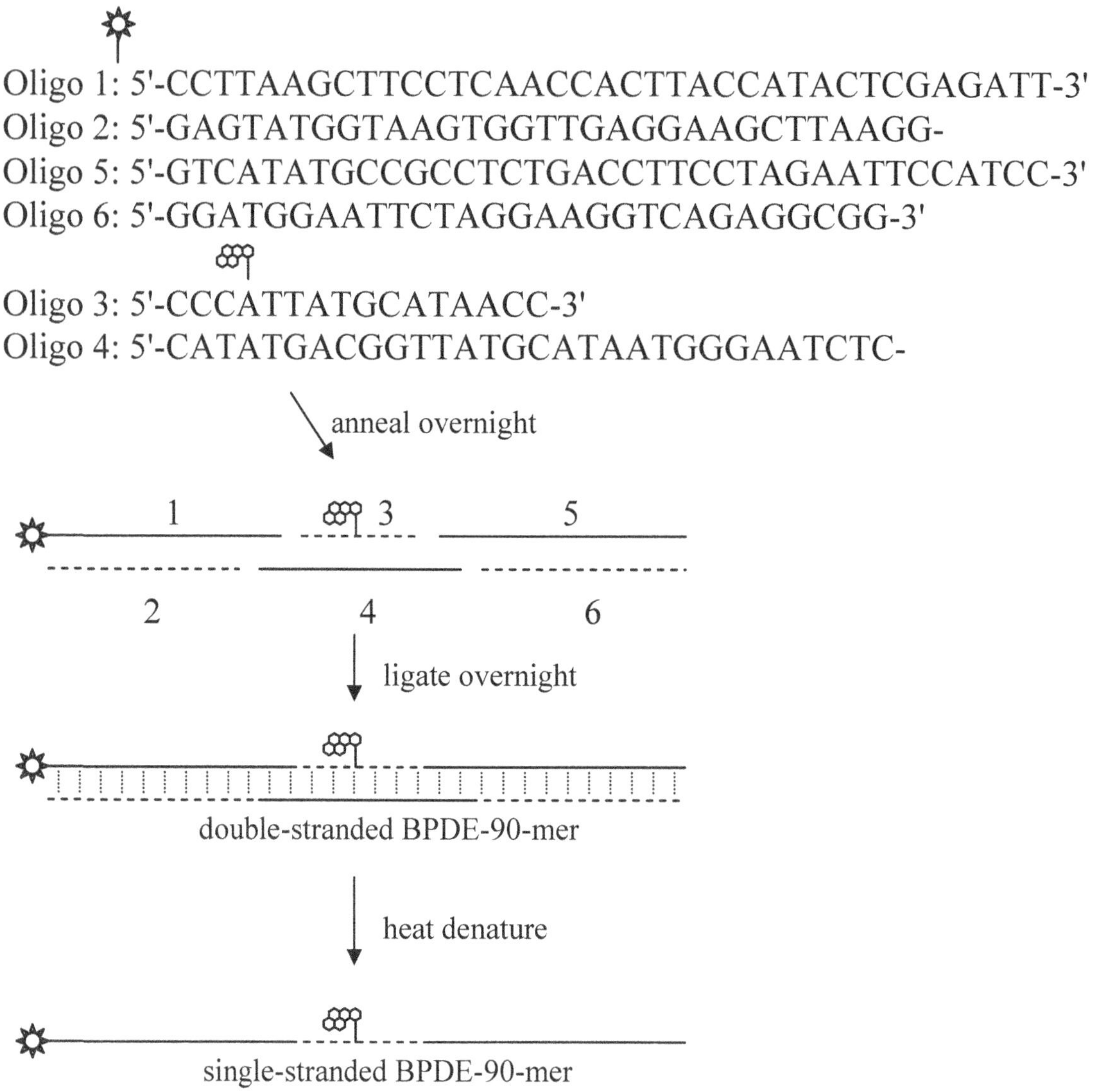


Figure 4.1. Schematic representation of the design of the DNA damage probe. A total of six overlapping, complementary oligonucleotides of varying lengths were annealed and ligated to form a 90-base pair double-stranded DNA molecule. Oligo 1 was labeled at the 5'-end with the fluorescent dye tetramethylrhodamine (TMR). Oligo 3 contained a damaged base, BPDE-$N^2$-deoxyguanosine, which was introduced by reacting oligo 3 with (±)-*anti*-BPDE. The dye label and damaged base were on the same strand to allow for either single- or double-stranded experiments. (*Source:* Carnelley et al. 2001)

## Synthesis of Damaged Oligonucleotide

(±)-*anti*-BPDE was used as the model carcinogen for synthesis of the damaged oligonucleotide. A 16-mer with the sequence 5'-CCCATTATGCATAACC-3' was synthesized to optimize the yield of the BPDE-$N^2$ deoxyguanosine (dG) adduct (Margulis, Ibanez, and Geactinov 1993; Funk et al. 1997). The TGC sequence is known to favor the formation of BPDE adduct. The BPDE-oligonucleotide reaction was based on the procedure described by Margulis, Ibanez, and Geactinov (1993), with slight modifications. The 16-mer was diluted in 20 mM phosphate buffer, (pH 11), containing 1.5% triethylamine, to a concentration of 60 μM in a volume of 400 μL. A fresh 3 mM solution of (±)-*anti*-BPDE in DMSO was prepared, and 40 μL was added to the oligonucleotide solution. This corresponded to a BPDE: oligonucleotide ratio of 5:1. The reaction was carried out at room temperature for 20 hours in the dark with gentle shaking.

## Purification of BPDE-Oligonucleotide

The components in the BPDE-oligonucleotide reaction mixture were separated using reversed-phase HPLC. The HPLC system consisted of a Dionex (Sunnyvale, Calif.) AGP1 advanced gradient pump with online-degassing module, either an analytical or preparative C18 column, and a Waters (Milford, Mass.) 484 tunable absorbance detector in series with a Shimadzu RF-551 fluorescence HPLC monitor (Columbia, Md.). The detectors were connected to a Hewlett Packard Model 35900 multichannel interface (Palo Alto, Calif.), which converted the signals for use by a computer running ChemStation software (Hewlett Packard, Palo Alto, Calif.). Preparative separation was carried out on a 10.0 X 250 mm, 5 μm Luna C18(2) preparative column (Phenomenex, Torrance, Calif.). The reaction products were initially assessed on the analytical column using a protocol described previously (Margulis, Ibanez, and Geactinov 1993; Cosman et al. 1990). This procedure employed a linear 0-90% methanol gradient in 20 mM sodium phosphate buffer (pH 7.0) in 60 min, with a flow rate of 0.75 mL/min. To reduce separation times for large volumes of the reaction mixture, HPLC purification of the BPDE-16-mer was carried out in two steps. The first separation was under isocratic conditions, using a mobile phase of 70% methanol / 30% 20 mM sodium phosphate, pH 7.0 buffer and a flow rate of 0.75 mL/min and 3.5 mL/min for the analytical and preparative columns, respectively. Elution of products were monitored in series by the absorbance detector (wavelength = 260 nm for DNA) and the fluorescence detector (excitation wavelength = 343 nm, emission wavelength = 400 nm for BPDE). This first separation removed unreacted BPDE as well as the tetrol hydrolysis products. DNA fractions were collected, dried using a centrifugal evaporator, and redissolved in distilled deionized water ($ddH_2O$). The second separation consisted of a linear 10-40% methanol / 20 mM sodium phosphate, pH 7.0 buffer gradient in 7.5 min (4%/min) followed by an additional 5 minutes at 40% methanol. This separated the BPDE-oligonucleotide from unreacted oligonucleotide. BPDE-oligonucleotide fractions were collected, dried to remove methanol and redissolved in $ddH_2O$. The samples were desalted using Sep-Pak C18 reversed-phase columns (Waters).

The sample was applied to a prepared Sep-Pak cartridge, then washed with 10 mL of the following solutions: 25 mM ammonium bicarbonate (pH 8.0); 25 mM ammonium bicarbonate / 5% acetonitrile; $H_2O$ / 5% acetonitrile; $H_2O$ / 5% acetonitrile. The BPDE-oligonucleotide was then eluted with 4 X 1 mL of $H_2O$ / 30% acetonitrile, dried and redissolved in $ddH_2O$.

**Synthesis and Purification of 90-mer Oligonucleotides**

Prior to ligation with the other 5 oligonucleotides, it was necessary to phosphorylate the freshly purified BPDE-16-mer at the 5'-end. Reaction mixtures included: ~200 pmol of BPDE-16-mer or control 16-mer, 4 μL of 100 μM ATP (400 pmol), 1.2 μL of 10X polynucleotide kinase reaction buffer, and $ddH_2O$ to a total volume of 12 μL. T4 polynucleotide kinase (PNK) was added (1 μL, 6.1 units/μL), then samples were mixed and incubated at 37°C for 1 hour. After complete reaction, the excess PNK was heat denatured at 70°C for 10 minutes. The 16-mers were then mixed with the TMR-labeled 37-mer and the other 4 oligonucleotides so that all would be in 2:1 excess over the 16-mers. 5X DNA ligase buffer was added to a final concentration of 1X and the mixture was heated in a water bath to 70°C for 10 minutes, then allowed to cool over several hours to room temperature. DNA ligase was added (2 μL, 8.5 Weiss units/μL) and the sample incubated overnight at 16°C.

Purification of the BPDE and control ligation products was achieved using preparative, 7.5% native polyacrylamide gel electrophoresis (PAGE). Electrophoresis was carried out at 600 V for 6 hours with a water cooling core to prevent denaturation of the ligation products. The bands were visualized by brief exposure to ultraviolet light, causing the TMR label to fluoresce, and cut from the gel. The gel slices were crushed and soaked to elute the products overnight in 0.3 M sodium acetate, pH 5.2 on a rotary shaker protected from light. After elution, polyacrylamide fragments were removed from solution using filter units prepared in the lab. The solution was passed through silanized glass wool followed by GF/C glass microfibre filter paper (Whatman). The samples were then extracted and back-extracted with equal volumes of phenol/chloroform/isoamyl alcohol (25:24:1) followed by chloroform/isoamyl alcohol (24:1). Oligonucleotides were precipitated by adding $MgCl_2$ to 10 mM and 3 volumes of ice-cold 95% ethanol and, then placed at -20°C overnight. The following day samples were centrifuged for 45 minutes at 14000 rpm and 4°C, supernatant was removed, and the pellets were washed once with 95% ethanol. Samples were again centrifuged for 10 minutes, dried and redissolved in $ddH_2O$. UV-Vis absorbance scans were performed on the resulting oligonucleotide solutions to determine concentration as well as to confirm the presence of the TMR dye and BPDE moiety.

**Instrumentation for Analysis of Ligation Products**

Analysis and characterization of the BPDE and control ligation products was carried out using a laboratory-built capillary electrophoresis laser induced fluorescence (CE/LIF) system as previously described (Le et al. 1995; Wan and Le 2000) (Figure 4.2). Electrophoresis was powered by a high voltage power supply (CZE1000R, Spellman High Voltage Electronics, Plainview, NY). Separation conditions including sample injection time and voltage, separation voltage and run time were controlled by LabVIEW (National Instruments, Austin, Texas) program run on a Macintosh computer. Capillaries used for these experiments were uncoated fused silica (Polymicro Technologies, Phoenix, Ariz.), with a 50 μm i.d., 150 μm o.d., 42 cm total length, and 37 cm effective separation length. The injection end of the capillary was placed in sample solution or running buffer, along with the high voltage lead from the power supply. The other end of the capillary was inserted through a grounded holder and into a waste vial.

The laser-induced fluorescence detector was built on an optical table using both commercial equipment and custom-made accessories. The laser source was a 1.0 mW green helium-neon laser (Melles Griot, Irvine, Calif.) with an excitation wavelength of 543.5 nm. The laser was focused onto the capillary through a microscope objective (6.3x), and fluorescence was collected through a second, high numerical aperture objective (60x, 0.7 NA, Universe Kogaku, Oyster Bay, N.Y.) positioned at $90^o$ from the direction of the laser. The fluorescence signal passed through a pinhole with an adjustable diameter to minimize background light, and also through a bandpass filter (580DF40) to eliminate scattered laser light. The signal was detected by a photomultiplier tube (R1477, Hamamatsu Photonics, Japan), and recorded by a Macintosh computer running LabVIEW software and equipped with a PCI data acquisition board.

The system was equipped with an auxiliary microscope to assist in the alignment of the optics. The microscope was used to visualize the position of the laser beam with respect to both the sample flow through the capillary and the collection optics, represented by a light-emitting diode (LED) positioned behind the pinhole in the collection assembly. Alignment was achieved by initially fixing the position of the collection assembly, then adjusting the capillary and laser-focusing objective using X-Y-Z translation stages. The angle of the fluorescence-collecting objective and the position of the collection assembly were also adjustable for optimization of alignment.

Samples were electrokinetically injected into the capillary by applying an injection voltage of 10000 V for 5 seconds. The separation was carried out at room temperature with a separation voltage of 20000 V. The running buffer used was 1X Tris-glycine (25 mM Tris, 250 mM glycine), pH 8.3. The capillary was washed approximately every 5-10 injections with 0.1 M NaOH (applied by syringe for 1 min) followed by electrophoresis using 1X Tris-glycine, pH 8.3 for 7 minutes. The initial voltage was kept low to prevent excessive joule heating in the capillary. As the running buffer replaced the NaOH in the capillary, current decreased allowing the running voltage to be gradually increased to 20000 V for the final 5 minutes of the reconditioning period. All capillary electrophoresis data were analyzed using Igor Pro software (version 3.1, WaveMetrics Inc., Lake Oswego, Ore.).

**Characterization of BPDE and Control 90-mers**

Prior to analysis, 90-mer samples were diluted to appropriate concentrations in running buffer (1X Tris-glycine, pH 8.3). The 90-mer products were analyzed either in their native form or their denatured, single-stranded form. Denaturation of the 90-mers was achieved by heating the samples at $100^oC$ for 10 minutes in a heating block, then transferring directly to ice to prevent reannealing. After cooling, the samples were briefly centrifuged in a microcentrifuge to collect condensation from the side of the tube, then gently mixed to ensure a homogenous solution. Total sample volume was typically 20 μL, which allowed for convenient injection into the capillary. For experiments involving antibodies, fresh dilutions of antibody stock solutions were prepared immediately before analysis and kept on ice. After addition of antibody to the 90-mer solution, the sample was gently vortexed to ensure complete mixing.

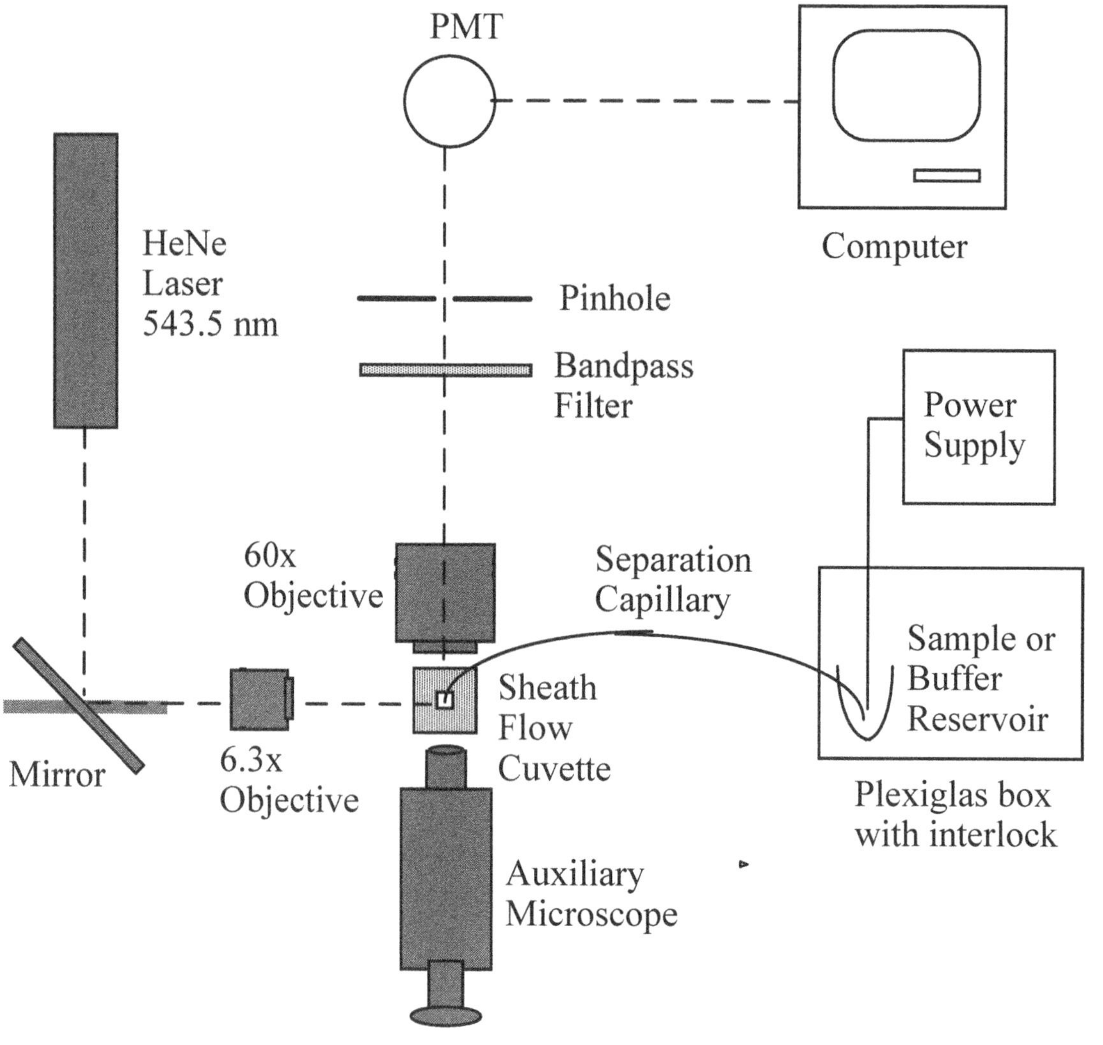


Figure 4.2. Schematic showing capillary electrophoresis with laser-induced fluorescence detection system.

**Treatment of A549 Cells with BPDE**

A human lung carcinoma cell line (A549) was incubated with BPDE to produce DNA adducts in genomic DNA. The cell line was maintained in DMEM/F12 medium (Gibco BRL, Gaithersburg, Md.) supplemented with 10% fetal bovine serum. The cells were seeded at 1 x $10^5$ cells per plate and maintained at 95% humidity and 5% $CO_2$ for 20 hours prior to the addition of BPDE. Old culture media were removed from each culture plate and the cells were washed twice with phosphate buffered saline (PBS). Media containing BPDE at various concentrations (0, 2.5, 5, and 10 μM final concentration) were added to the designated plates. The cells were further incubated in the media containing BPDE for 2 hours. The cells were then washed with PBS prior to the addition of DNAzol lysis reagent (Gibco BRL) to facilitate cell lysis and DNA extraction. Subsequent steps involved a 99.9% ice cold ethanol precipitation and a 70% cold ethanol wash to purify the genomic DNA. The final DNA pellet was dissolved in distilled deionized water ($ddH_2O$) and DNA concentration was measured at $OD_{260}$ using $ddH_2O$ as a blank.

**Competitive Assay for BPDE-DNA Adducts**

The DNA samples from the A549 cells were analyzed for BPDE-DNA adducts by competitive assay using the TMR-labeled 16-mer or 90-mer oligonucleotides as probes. Mixtures containing 60 nM of the oligonucleotide probe, 0.4 μg/mL of mouse monoclonal antibody 8E11, and 80 μg/mL of the DNA from A549 cells were incubated in 20 μL of tris-glycine buffer (25 mM tris and 200 mM glycine, pH 8.3) at room temperature for 30 min. These were subjected to CE/LIF analysis to detect both antibody-bound and unbound fluorescent probes.

## RESULTS AND DISCUSSION

**Purification of BPDE-16-mer Oligonucleotide**

The reaction of (±)-*anti*-BPDE with the 16-mer resulted in a mixture of products including unreacted 16-mer, BPDE-modified 16-mer, and numerous BPDE hydrolysis products. These reaction products were separated by reversed-phase HPLC. The initial conditions used were essentially identical to those used by Margulis, Ibanez, and Geactinov (1993). It took 52 min to separate the modified (±)-*anti*-BPDE-$N^2$-dG 16-mer adducts from the unreacted 16-mer and the hydrolysis byproducts of unreacted BPDE in solution.

To reduce the time and volume of mobile phase required to purify large amounts of the BPDE-16-mer, the purification protocol was revised using two separation steps. The first step was intended to separate the unreacted BPDE and hydrolysis products from the BPDE-16-mer and unmodified 16-mer. Because of the large difference in hydrophobicity of BPDE with the oligo species, they were easily separated over a relatively short period of time.

Figure 4.3 shows typical chromatograms from this first separation using the analytical column. The three major species eluted from the column with retention times of 2.24, 6.25, and 7.26 min. The first peak (2.24 min) corresponds to a mixture of unmodified 16-mer and BPDE-16-mer, and it has both the strong UV absorbance and the fluorescence signal from the BPDE. For the control sample containing only 16-mer (no BPDE), the retention time of the DNA peak was identical, but there was no signal in the fluorescence trace. The two later peaks in Figure 4.3

(6.25 ad 7.26 min) correspond to the tetrol hydrolysis products from BPDE. These peaks, as well as small peaks between 3 and 6 minutes, were identical in the control sample containing only reaction buffer and BPDE (no oligonucleotide). When these samples were applied to the preparative column the separation was very similar, with retention times of the peaks described above varying by only a few seconds. The main difference between the analytical and preparative columns was that the injection volume was scaled up significantly for preparative separation. This resulted in a much larger signal from the two detectors, without significant peak broadening.

The second step in the purification method was to separate the mixture of BPDE-16-mer and unmodified 16-mer collected in DNA fractions from the first step. This step utilized a linear gradient of methanol in sodium phosphate buffer (10-40%, 4%/min) to improve the separation. A typical run using the analytical column is shown in Figure 4.4. The difference between the BPDE-16 mer and unmodified oligo was clearly demonstrated based on their differences in absorbance and fluorescence characteristics. The peak eluting at 12.01 min in the UV trace represented the unmodified 16-mer, with no fluorescence detected. The doublet peaks with maxima at 13.07 and 13.45 min represent a mixture of two stereoisomers of the BPDE-16-mer reaction product. These different isomers have been described previously (Cosman et al. 1990). For our purpose, the isomers were pooled together to generate sufficient starting material for the ligation procedure and to more fully represent the characteristics of DNA damaged by (±)-*anti*-BPDE. When the separation was scaled up using the preparative column, the results were very similar retention times and peak shapes, with a higher signal corresponding to the larger amount of sample injected. The purified BPDE-16-mer was collected in fractions from the preparative column, dried and redissolved for use in the ligation reaction. A UV-visible wavelength scan of the BPDE-16mer showed maxima of 260 nm and 352 nm, corresponding to absorbance by DNA and BPDE, respectively.

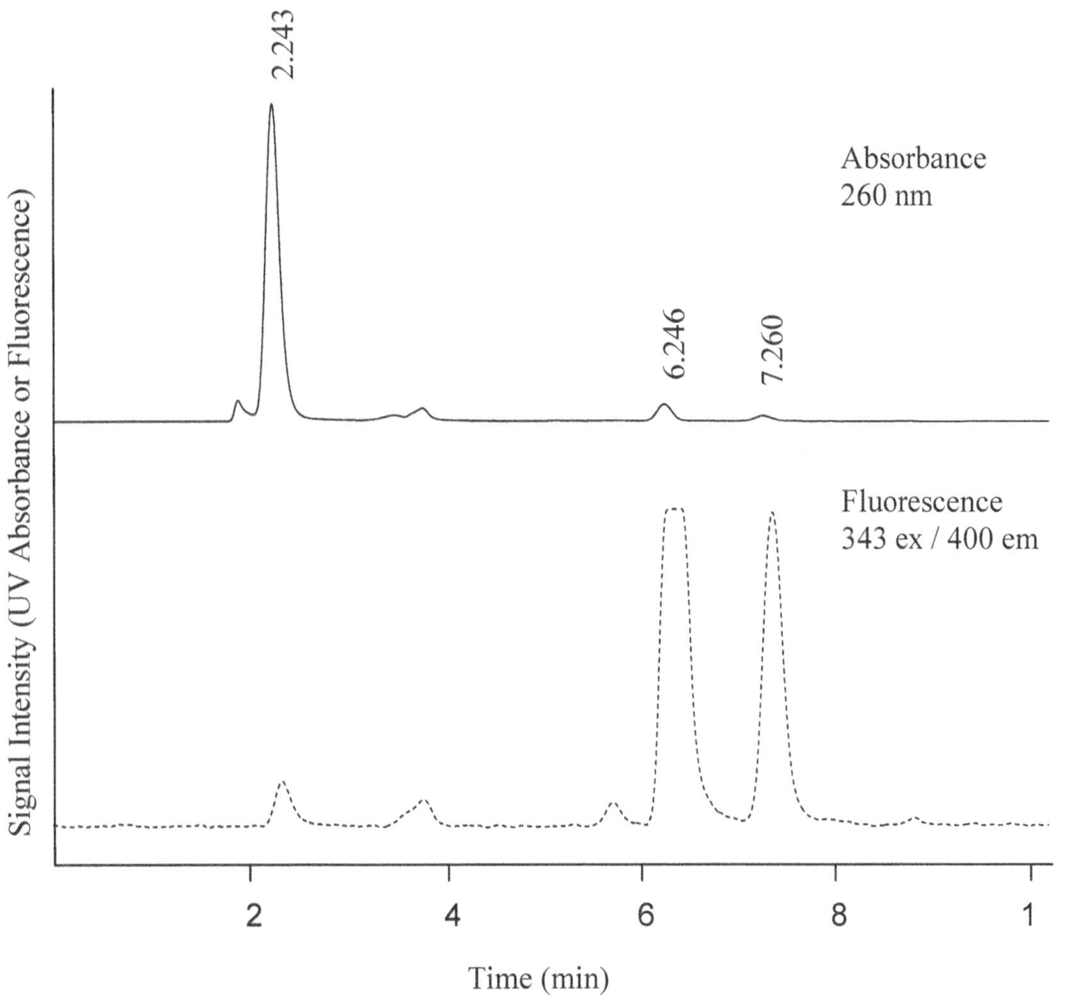


Figure 4.3. First step in the HPLC purification of BPDE-modified 16-mers from a reaction involving (±)-*anti*-BPDE and a 16-mer oligonucleotide. Removal of BPDE hydrolysis products was achieved using an isocratic mobile phase of 70% methanol / 30% 20 mM sodium phosphate buffer (pH 7). Major species eluted from the column were: a mixture of unmodified and BPDE-modified 16-mers (2.24 min); BPDE tetrol hydrolysis products (6.25 and 7.26 min). The 16-mer fraction was collected and subjected to a second purification step. (*Source:* Carnelley et al. 2001)

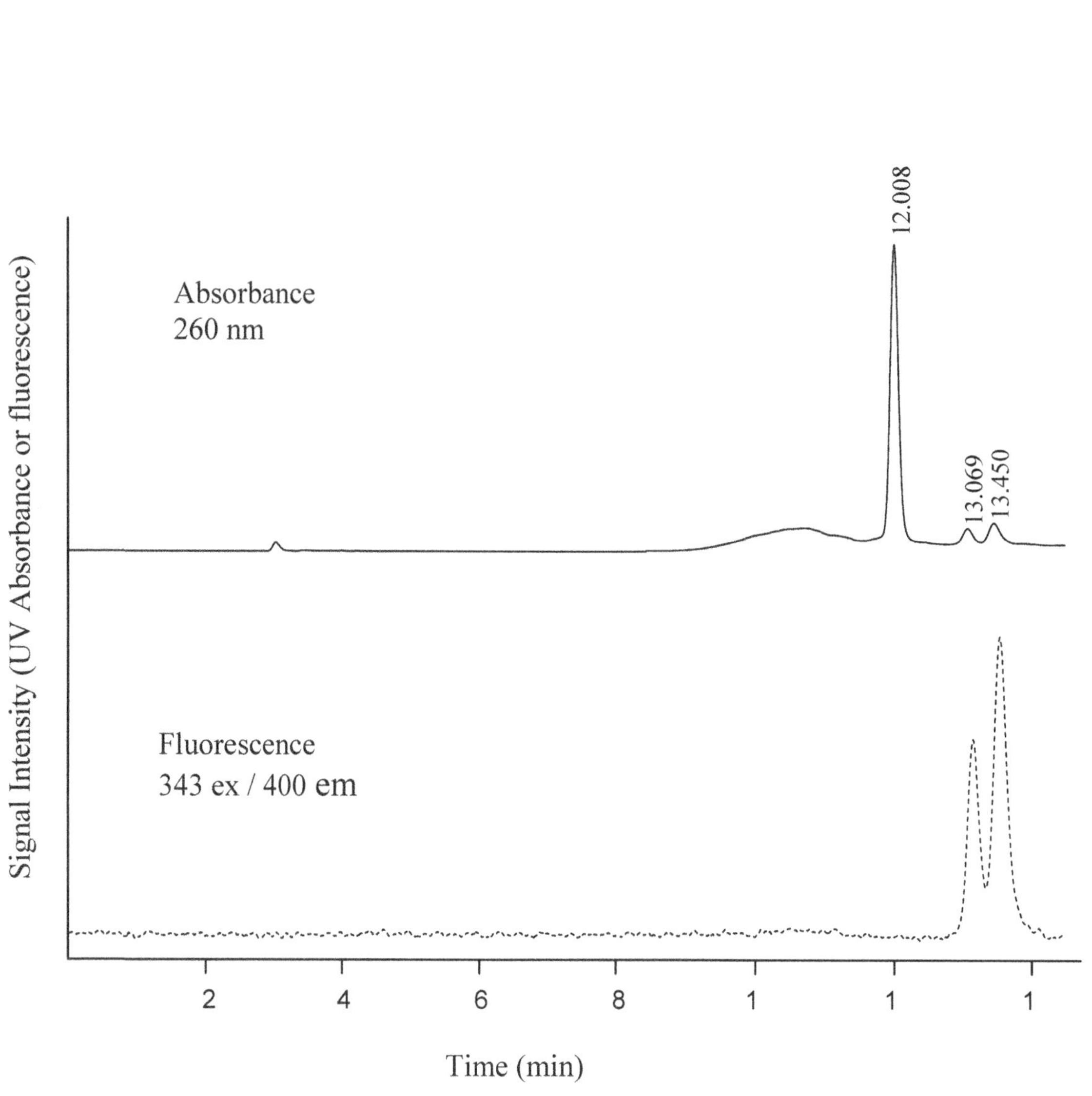


Figure 4.4. Second step in the HPLC purification of BPDE-modified 16-mers. The oligo species in the DNA fraction collected during the first step were separated using a linear 10-40% methanol gradient in 20 mM sodium phosphate buffer (pH 7) in 7.5 min (4%/min), followed by an additional 5 min at 40% methanol. Species eluted from the column were: unmodified 16-mer (12.01 min); BPDE-modified 16-mer stereoisomers (13.07 and 13.45 min). (*Source:* Carnelley et al. 2001)

## Synthesis and Purification of BPDE 90-mer Ligation Products

Ligation reactions involving the six different oligonucleotides were carried out using the BPDE-16-mer purified by HPLC as well as a control 16-mer with the same sequence but without the BPDE modification. The products were run on a preparative native-PAGE gel to identify and recover the full length, double-stranded ligation products. Both the BPDE and control lanes contained a number of different bands that were visualized by TMR fluorescence after brief exposure to ultraviolet light. A photograph of the gel was not taken, in order to avoid both photobleaching of the TMR dye as well as UV-induced damage to the 90-mer construct. In both cases there was a major band corresponding to the full-length product, several minor bands of lower molecular weight indicating partial ligation products, and excess TMR-labeled 37-mer. The 37-mer was identified by running it alone in an adjacent lane on the gel. These extra bands were expected in the reaction as the TMR-labeled 37-mer and the other four oligos were in 2:1 excess over the BPDE- or control-16-mer.

The full-length BPDE and control 90-mers were recovered from the gel, filtered, purified by phenol-chloroform extraction and ethanol precipitation, and redissolved for further analysis.

## Characterization of the BPDE 90-mer Ligation Products

Both the BPDE-90-mer and control 90-mer were subjected to a UV-visible wavelength scan to confirm the nature of the modifications. The control 90-mer showed absorbance maxima at 260 nm and 556 nm. These corresponded to the absorbance properties of DNA and TMR, respectively. The BPDE-90-mer also demonstrated absorbance maxima at 260 nm and 556 nm, as well as a maximum at 347 nm that was not present in the scan from the control 90-mer. This wavelength was slightly different from the maximum measured for the BPDE-16-mer (352 nm). However, this slight shift in the BPDE absorbance was likely due to the oligonucleotide being double-stranded instead of single-stranded. These wavelength scans support the incorporation of BPDE into the BPDE-90-mer and the absence of this modification in the control 90-mer.

Further characterization of the BPDE and control 90-mers was carried out using capillary electrophoresis with laser-induced fluorescence detection. Both products were injected in either their native or denatured form, resulting in electropherograms shown in Figure 4.5. The two 90-mers were similar in their behavior in CE. In the native samples, there was a doublet peak migrating between 4.0 and 4.5 minutes. The ratio of the two peaks varied slightly between runs for a given 90-mer, but the doublet was stable and reproducible. When the 90-mers were denatured by heating at 100°C for 10 minutes before injection, the result was a single peak migrating at 4.1 min and 4.2 min for the BPDE and control, respectively. For both 90-mers the migration of the denatured peak was very similar to the first of the two peaks in the doublet observed for the native samples. After heat denaturation the 90-mer existed entirely in the single-stranded form, resulting in a single peak in the electropherogram.

It was observed (Figure 4.5) that the single peak achieved a higher fluorescent intensity (peak height) than the doublet peak. Furthermore, the total area of the single peak was significantly higher than the total area of the doublet peak for identical injections from the same sample before and after heat denaturation. This occurred for both the BPDE and control 90-mers, and was typically an increase of 120-130% in the total fluorescent signal. This difference was likely a result of the TMR fluorescence being quenched when in the double-stranded oligonucleotide complex. Upon denaturation to the single-stranded form, the TMR dye recovered its full fluorescence yield. This effect has been demonstrated previously in melting curve experiments (Vamosi and Clegg 1998). The overall quantum yield, fluorescence lifetime, and fluorescence intensity of TMR-labeled oligonucleotides ranging in length from 8 to 34 base pairs increased significantly during the transition from double-stranded to single-stranded form. Furthermore, it has been shown that TMR is selectively quenched by guanine bases in DNA (Seidel, Schulz, and Sauer 1996), with photo-induced electron transfer from TMR to guanine resulting in lower fluorescence lifetimes and quantum yields (Eggeling et al. 1998). This has been demonstrated by measuring fluorescence lifetimes of individual TMR-labeled oligonucleotide molecules (Eggeling et al. 1998; Edman, Mets, and Rigler 1996). Although there are no guanines in the immediate vicinity of the TMR label at the 5'-end of oligo 1, the complementary oligonucleotide (oligo 2) has two guanine residues at its 3'-terminus. In the double-stranded form these two bases are directly adjacent to the TMR dye, and are the likely reason for the observed quenching of the fluorescent signal.

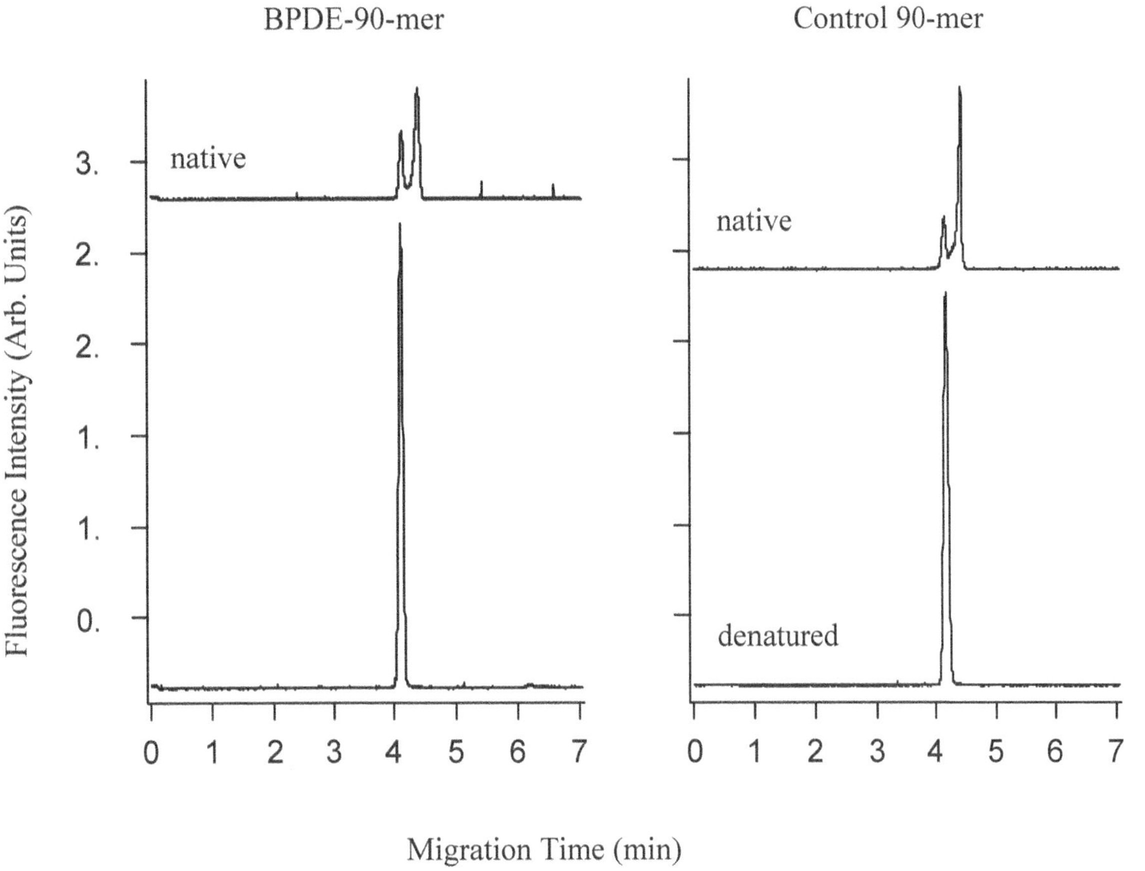


Figure 4.5. Effect of heat denaturation on BPDE-modified and control 90-mers. In the native state, both 90-mers exist as a mixture of single- and double-stranded DNA during capillary electrophoresis (doublet peak). After denaturation, both 90-mers are completely in the single-stranded form (single peak). Capillary electrophoresis experiments were carried out using bare fused silica capillaries (50 μm i.d., 37 cm effective length), 1X Tris-glycine running buffer (pH 8.3), and a running voltage of 20000 V. The 90-mer samples ($5x10^{-9}$ M) were electrokinetically injected into the capillary by applying an injection voltage of 10000 V for 5 sec. Fluorescence was excited at 543.5 nm with a green HeNe laser and detected at 580 nm. (*Source:* Carnelley et al. 2001)

**Affinity Interaction of BPDE-90-mers with a Monoclonal Antibody**

Preliminary experiments using monoclonal antibody 8E11 demonstrated that the specific antibody bound to the BPDE-90 mer, not the control 90 mer (Figure 4.6). These results were obtained by first denaturing the 90-mers ($5x10^{-9}$ M), then adding 8E11 antibody to a final concentration of 20 μg/mL and incubating for 10 min at room temperature (21°C). The same fluorescence intensity scale was used for both 90-mers for ease of comparison. For the mixture of the BPDE 90-mer and 8E11, an additional peak was present in the electropherogram with a migration time of approximately 3.0 min. This peak represented the complex between the antibody and single-stranded BPDE-DNA, and was well-resolved from the denatured 90-mer peak at 4.1 min. When comparing the fluorescent signals between runs, the total area of the two peaks for the mixture of BPDE 90-mer and 8E11 was very similar to the area of the peak for the BPDE 90-mer alone. The formation of an antibody-DNA complex was not observed with the control 90-mer, indicating a specific interaction of the antibody with the BPDE 90-mer.

The effect of incubation time and temperature on complex formation were investigated using the same concentrations of BPDE 90-mer ($5x10^{-9}$ M) and 8E11 (20 μg/mL). For incubations carried out at both room temperature and on ice, the interaction did not change significantly between 1 min and 20 min. At room temperature, the complex was stable after 45 min. For incubation on ice, the complex decreased slightly after 45 min when compared to the 20 min incubation. In general, room temperature incubations with 8E11 resulted in more stable and reproducible complex formation than incubations on ice. This result is expected since the recommended temperature for conventional immunoassays using 8E11 is 37°C (Santella et al. 1984, Hsu et al. 1995), and most immunochemical procedures require incubation temperatures of either 37°C or room temperature (Harlow and Lane 1988). Based on these results, further experiments with 8E11 antibody were carried out at room temperature. An incubation time of 5 min was chosen for ease of sample preparation and analysis.

Overnight incubations resulted in a decrease of the DNA-antibody complex, as well as a reversion to the doublet shape for the free 90-mer peak as demonstrated in Figure 4.7. This result suggests that 90-mer samples left overnight tended to re-anneal to the double-stranded form, causing dissociation of the DNA-antibody complex. This also implies that the affinity of 8E11 for double-stranded BPDE-DNA is less than for single-stranded BPDE-DNA.

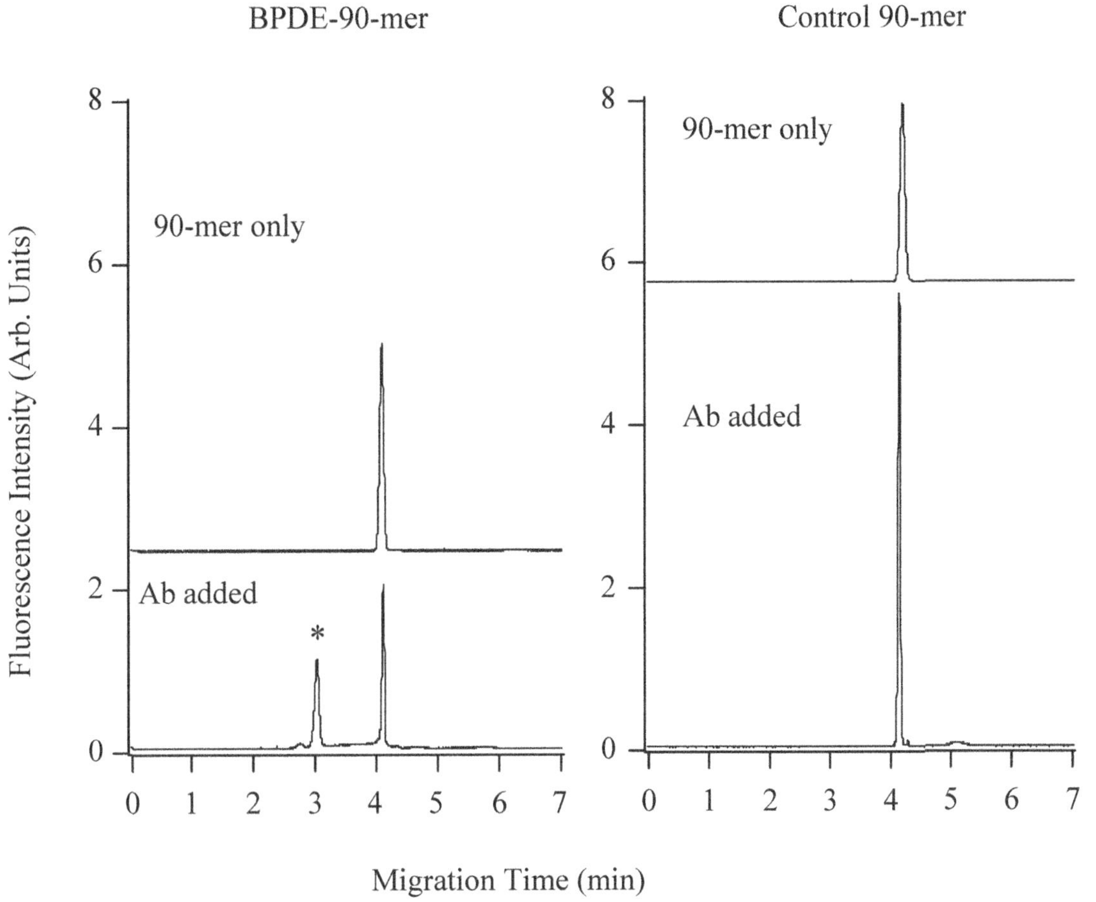


Figure 4.6. Capillary electrophoresis analysis of the fluorescent 90-mers and their mixtures with antibody 8E11. The addition of 20 μg/mL antibody 8E11 to heat denatured BPDE 90-mer ($5x10^{-9}$ M) resulted in the formation of a second peak corresponding to the DNA-antibody complex (3 min). This complex peak was not present for the control 90-mer. The same CE/LIF conditions as shown in Figure 4.5 were used. (*Source:* Carnelley et al. 2001)

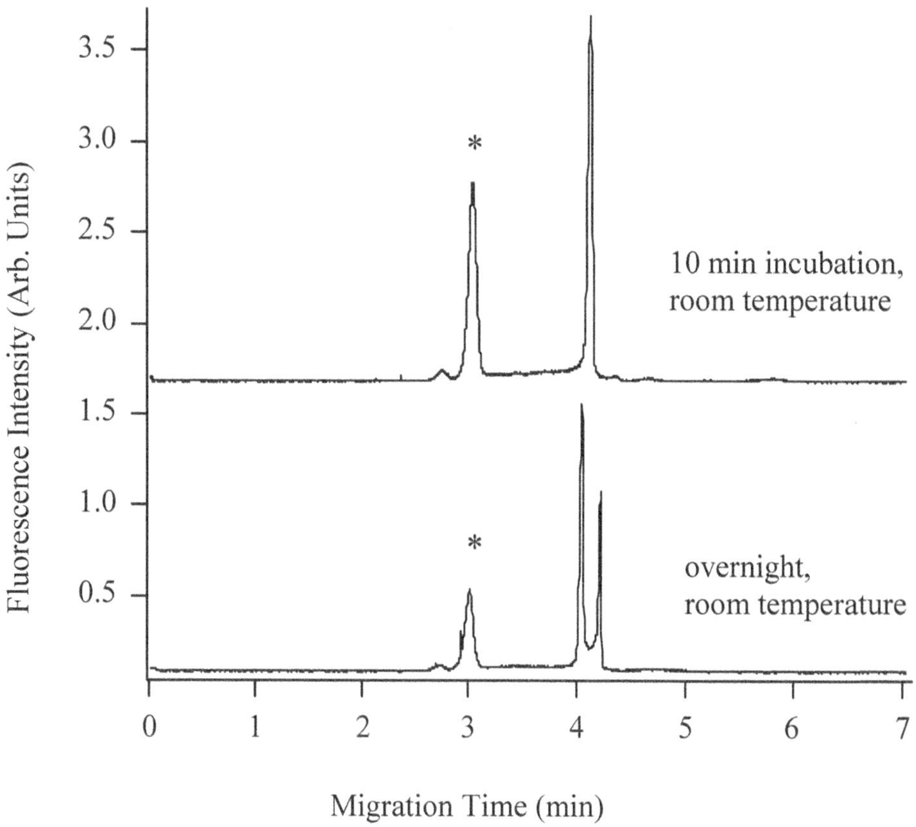


Figure 4.7. Effect of overnight incubation for BPDE 90-mer and 8E11 antibody. During the longer incubation time, the complex peak (marked with *) decreased and the free 90-mer peak reverted to a doublet shape. The same CE/LIF conditions as shown in Figure 4.5 were used. (*Source:* Carnelley et al. 2001)

The difference in affinity of 8E11 antibody between single- and double-stranded BPDE-modified DNA was further confirmed by comparing its binding with heat-denatured BPDE 90-mer and native BPDE-90 mer (20 μg/mL 8E11). The antibody-oligonucleotide complex formation was approximately 6 fold higher for the denatured single-stranded 90-mer than the native form (Table 4.1). Thus, the denaturation of samples by heat before incubation with antibodies was retained for further experiments.

Table 4.1.
Comparison of denatured and native BPDE-90-mers on their affinity with anti-BPDE antibody 8E11. Results from two independent trials for each form are shown.

| 90mer form | Antibody-90mer complex area* | Free 90-mer area* | Sum of areas* | Complex area / total area* |
|---|---|---|---|---|
| Denatured | 0.0993 | 0.0996 | 0.1989 | 50 |
| Denatured | 0.0966 | 0.0860 | 0.1826 | 53 |
| Native | 0.0084 | 0.0882 | 0.0966 | 8.7 |
| Native | 0.0083 | 0.0851 | 0.0934 | 8.9 |

*Source:* Carnelley et al. 2001

* Peak area values are in arbitrary units integrated by IgorPro data analysis software

**Determination of Specific Antibody Using BPDE-90 mer as a Probe**

An application of the fluorescent BPDE-90 mer probe was demonstrated for the determination of anti-BPDE antibody. Figure 4.8 shows a typical calibration from the analyses of mixtures containing different amounts of 8E11 and a constant concentration of the detatured BPDE-90 mer probe ($5 \times 10^{-9}$ M). A DNA-antibody complex peak was observed with 8E11 concentrations as low as 0.1 μg/mL (Figure 4.8, inset). This concentration corresponds to $0.7 \times 10^{-9}$ M (or 0.7 nM) assuming a molecular weight of approximately 150,000 for the antibody 8E11. The concentration of BPDE-90 mer (5 nM) was in excess and the formation of its complex with the antibody was not complete. The amount of the complex increased at higher concentrations of 8E11, up to 10 μg/ml (7 nM). At this concentration complex formation appeared to reach saturation, since further increase of antibody concentrations did not increase the proportion of 90-mer bound to 8E11 (Figure 4.9).

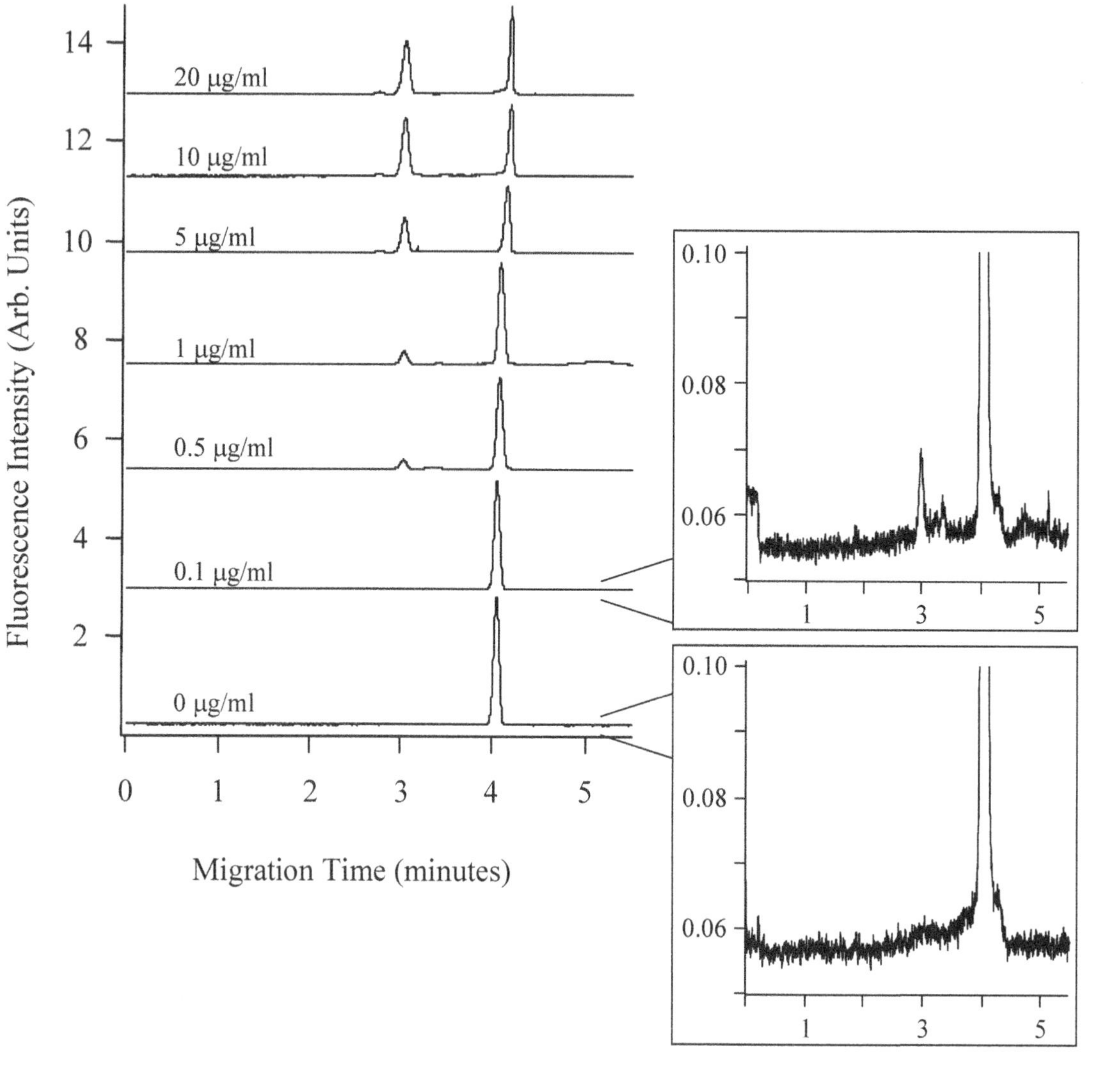


Figure 4.8. Incubation of BPDE 90-mer ($5x10^{-9}$ M) with varying concentrations of antibody 8E11. The BPDE 90-mer and antibody 8E11 were incubated at room temperature for 5 min prior to CE/LIF analysis. As the antibody concentration in the sample increased, the area of the complex peak increased until reaching a maximum at 10 μg/mL ($7x10^{-9}$ M). The peak at ~3 min correspond to the complex between the antibody and the BPDE 90 mer. Inset, right: expanded scale of 0.1 μg/mL 8E11 sample and the control sample. The same CE/LIF conditions as shown in Figure 4.5 were used. (*Source:* Carnelley et al. 2001)

## Screening for anti-BPDE Antibodies Using the Fluorescent BPDE-90mer Probe

The fluorescent BPDE-90 mer probe was further used to screen for specific binding proteins, with 3 antibodies as model protein analytes. Monoclonal antibodies 8E11, 5D11 and E5 are all specific for BPDE-modified DNA. A comparison between these antibodies was conducted to determine differences in their reactivity to the BPDE 90-mer standard as well as their behavior in the capillary electrophoresis system. Conditions used for sample preparation were identical to earlier experiments: heat denaturation of the 90-mer at 100°C for 10 min, cooling on ice, then incubation with antibody at room temperature for 5 min before injection. Polyclonal mouse IgG was used as a negative control since it is essentially the same molecular structure (isotype) as the monoclonal antibodies but is not expected to react with the BPDE 90-mer. The BPDE 90-mer probe concentration was fixed at $5 \times 10^{-9}$ M and the antibodies were added in varying amounts. All three monoclonal antibodies reacted with the 90-mer probe, with 8E11 giving the highest formation of complex (Figure 4.9). The negative control showed a very slight reactivity but was insignificant compared to the other antibodies, even at concentrations up to 40 μg/mL.

Antibodies 8E11 and E5 were found to bind specifically to the BPDE adduct. No cross-reactivity with the unmodified control 90-mer was observed for either 8E11 or E5. The antibody 5D11 showed slight cross-reaction with undamaged DNA. When incubated with 20 μg/mL 5D11, the control 90-mer formed a peak corresponding to antibody complex, with about 2.1% of total peak areas as compared with the BPDE 90-mer. This non-specific interaction between 5D11 and undamaged DNA is in agreement with previous studies (Santella et al. 1984) that have demonstrated cross-reactivity, and is a result of its being raised against a full-length BPDE-DNA antigen. Both 8E11 and E5 were raised against BPDE-guanosine monomers conjugated to carrier proteins (Baan et al. 1988, Santella et al. 1984) and therefore do not recognize undamaged DNA.

The incomplete binding of the DNA damage probe with the antibodies (up to 50% of binding) (Figure 4.9) is probably because the probe is a mixture of several BPDE-90 mer isomers. The stereochemistry of the BPDE-$N^2$-dG adduct could be important to its binding with specific antibodies. In the preparation of the BPDE-modified 16-mer, (±)-*anti*-BPDE was reacted with the oligonucleotide. The covalent bond that forms between BPDE and guanosine may be either *cis*- or *trans*- relative to the hydroxyl group on the adjacent carbon atom. Therefore, there may be as many as four different configurations of the BPDE 16-mer: (+)-*trans*, (+)-*cis*, (-)-*trans*, and (-)-*cis* (Szeliga and Dipple 1998). The reaction protocol was designed to minimize the formation of *cis*- adducts (Funk et al. 1997), but a mixture of (+)-*trans* and (-)-*trans* adducts with a small amount of *cis* adducts would be expected in the BPDE 16-mer reaction products (Cosman et al. 1990). Because these stereoisomers were pooled together after purification by HPLC and before the ligation reaction, the 90-mer product would also contain these configurations. The advantage of this mixture is that it more accurately represents the spectrum of damage that would occur in human DNA samples. The disadvantage is that BPDE-DNA antibodies exhibit different affinities for these stereoisomers (Hsu et al. 1995). In competitive inhibition studies using BPDE-modified 11-mers, Hsu et al. (1995) demonstrated a lower affinity for the (-)-*trans-anti*-BPDE-$N^2$-dG adduct than for the (+)-*trans-anti*-BPDE-$N^2$-dG adduct. For antibodies 8E11 and 5D11 this lower affinity was 66% and 20% of the (+)-*trans* adduct, respectively. Both antibodies exhibited much lower affinities for the *cis* adducts compared to the (-)-*trans* adduct. Since the 90-mer contained a combination of both *trans* adducts, the stereospecific difference in affinity may in part be responsible for the differences in complex formation observed for these antibodies. The presence of different BPDE-90mer isomers may also contribute to the observed incomplete binding. The other possible reason for the incomplete binding is the presence of residual oligonucleotides that do not contain BPDE and therefore, do not bind to the antibodies.

In addition to the isomer-specific reactivities, Hsu et al. (1995) showed a difference in affinity between 8E11 and 5D11 when considering only the (+)-*trans* adduct. 8E11 was approximately 7 times more sensitive than 5D11 for the very short 11-mer oligonucleotide. For full-length heat-denatured BPDE-DNA, the two antibodies were almost identical. This difference is likely due to the antigens against which these antibodies were raised: BPDE-$N^2$-dG mononucleotide for 8E11, full-length BPDE-DNA for 5D11. 5D11 may require a longer sequence of DNA surrounding the damaged site for binding which would not be present in the 11-mer. Given these results one might predict that for DNA of intermediate length (90 bases), 8E11 would still have a higher affinity than 5D11, but to a lesser extent. Our results (Figure 4.9) are consistent with these previous findings. These results indicate that monoclonal antibody 8E11 is likely the best choice for detecting BPDE-damaged DNA using the capillary electrophoresis / laser-induced fluorescence assay.

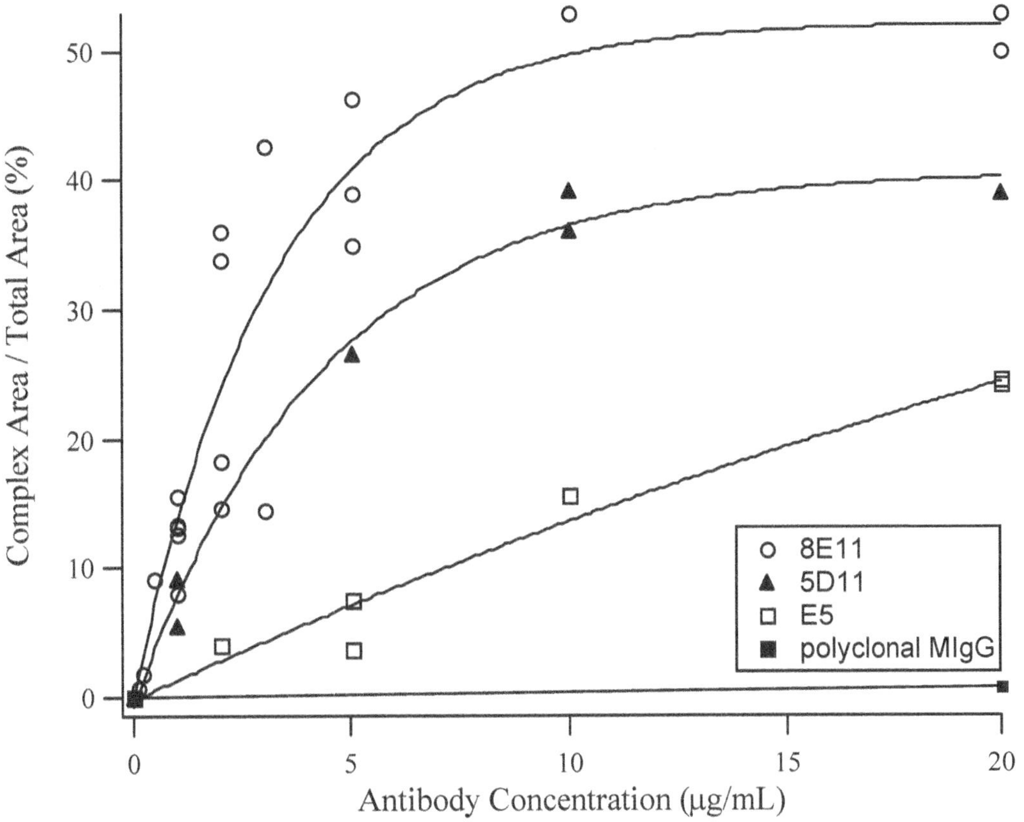


Figure 4.9. Comparison between BPDE-DNA antibodies using the BPDE 90-mer fluorescent probe. $5x10^{-9}$ M 90-mer was incubated with antibody for 5 minutes at room temperature and subjected to capillary electrophoresis. Polyclonal mouse IgG (MIgG) was the negative control. Error bars indicate the standard deviation of complex formation for antibody concentrations with replicate samples. The same CE/LIF conditions as shown in Figure 4.5 were used. (*Source:* Carnelly et al. 2001)

**Application of the BPDE-DNA Probe to Competitive Assay for BPDE-DNA Adducts in Cells**

The 90-mer probe described in this paper has many potential uses in DNA damage research. It enables the investigation of alternative assay methods, including CE-based competitive immunoassays (Tao and Kennedy 1996, Ye et al. 1998, Lam et al. 1999, Wan and Le 1999) using the probe as a fluorescent probe (competitor). This approach is based on competition between damaged DNA and the fluorescent probe for the binding sites of a limited amount of antibody. With little or no damaged DNA in a sample, the probe achieves maximum complex formation with the antibody. As the amount of damaged DNA in the sample mixture increases, the probe is displaced from the antibody. This would result in an increase in the free probe peak and a decrease in the probe-antibody complex peak. This method has been demonstrated by using oligonucleotide and genomic DNA containing BPDE-damaged sites (Tan et al. 2001). Figure 4.9 shows typical electropherograms from the analysis of BPDE-DNA adducts in A549 cells that were incubated with 2.5, 5, and 10 μM BPDE for 2 hr. Figure 4.9 shows that increasing amounts of BPDE-DNA adducts were formed as the cells were incubated with increasing concentrations of BPDE. The BPDE-DNA adducts compete with the TMR labeled BPDE-DNA adduct probe for the antibody binding, resulting in the corresponding increase of the unbound probe (peak 3) and decrease of antibody complexes (peaks 1 and 2) of the fluorescent probe. This analysis requires less than 4 min per separation and has excellent resolving power to separate the bound and unbound DNA adducts. The same approach may be extended to assays for other types of DNA damage.

Because the commonly used antibody IgG is bidentate, it is able to bind to two antigen molecules. Figure 4.10 shows that two complexes (peaks 1 and 2) are formed between the antibody 8E11 and the fluorescent probe. Complexes peak 1 and 2 probably correspond to the 1:1 (binary) and 1:2 (tertiary) stoichiometry, respectively. Thus, the fluorescent probe is also useful for studies of affinity binding stoichiometry (Wang et al. 2002).

Another important aspect of the probe's design is the flexibility to substitute different damage types in the molecule with relative ease. The sequences of the two center oligonucleotides may be changed depending on the desired modification. By inserting these different damaged oligos, a variety of DNA damage detection systems can be investigated using the corresponding damage probe and CE/LIF. The technique itself combines specific recognition with high sensitivity detection, minimal sample preparation, and fast analysis times (5 minutes per run).

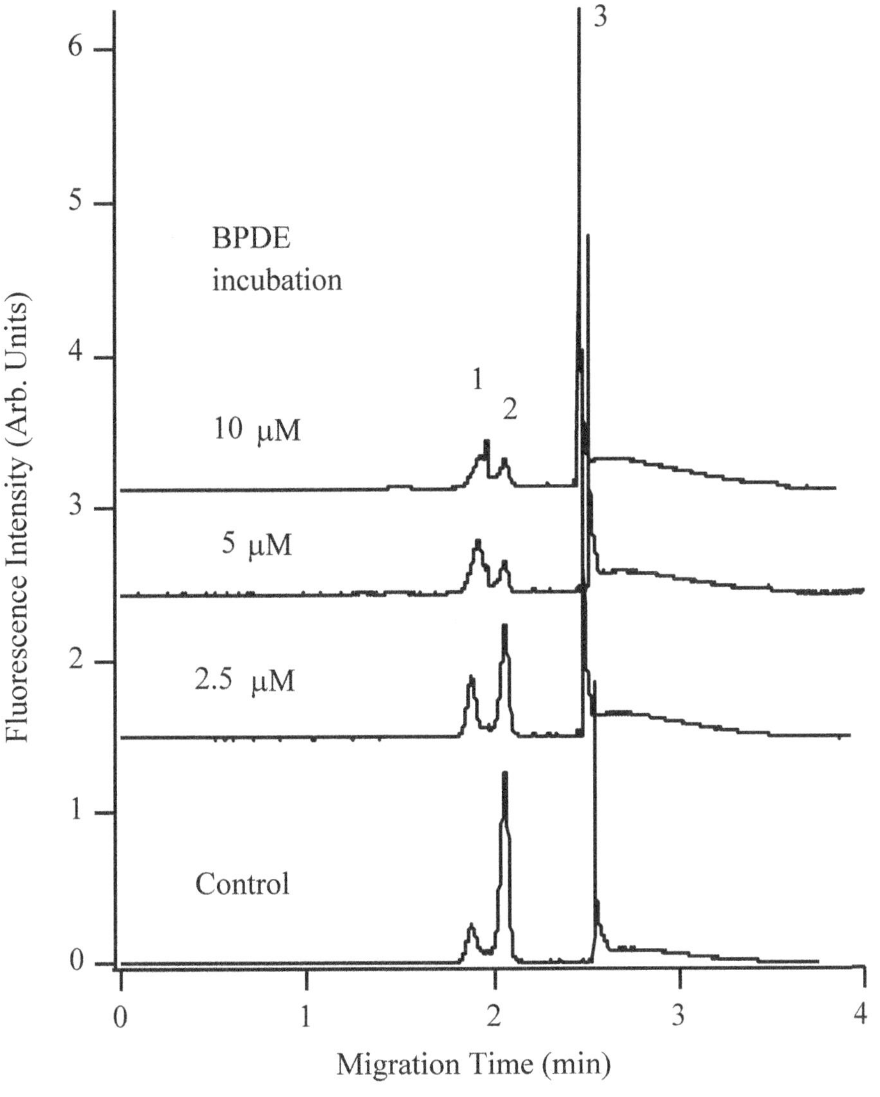


Figure 4.10. Representative electropherograms showing competitive assay for BPDE-DNA adducts from A549 cells. A549 cells were incubated with 0, 2.5, 5, and 10 μM BPDE for 2 h and DNA extracted for analysis. Mixtures containing 80 μg/mL of the cellular DNA, 60 nM of the oligonucleotide probe, and 0.4 μg/mL of mouse monoclonal antibody 8E11 were incubated in 20 μL of tris-glycine buffer (25 mM tris and 200 mM glycine, pH 8.3) at room temperature for 30 min. CE/LIF analysis of these mixtures were carried out using a fused silica capillary (30 cm in length, 20 μm i.d. and 150 μm o.d.) for separation. The separation buffer contained 25 mM tris and 200 mM glycine (pH 8.5). The running voltage was 30 kV and electrokinetic injection was carried out at 10 kV for 10 sec. Peaks 1 and 2 correspond to the binary (1:1) and tertiary (1:2) complexes between the antibody and the probe. Peak 3 represents the unbound probe. (*Source:* Carnelley et al. 2001)

# CHAPTER 5
# EFFECTS OF ARSENITE AND BPDE IN THE INDUCTION AND REPAIR OF BPDE-DNA ADDUCTS IN A549 CELLS.

## INTRODUCTION

The competitive assay described in Chapter 4 used an antibody and a labeled probe to determine unlabeled DNA adduct in cell samples. Competitive assays are usually limited by background and are less sensitive than direct assays of non-competitive format. This chapter describes the development of a non-competitive assay and its application to the determination of BPDE-DNA adducts in cells. The non-competitive immunoassay uses two antibodies, one of which is labelled fluorescent secondary antibody and the other is unlabeled primary antibody 8E11 that is specific for BPDE damage. Immunoassays measure damage more directly and do not require digestion of the cellular DNA, thereby reducing any artifacts that may be produced during DNA digestion and treatment procedures. The assay is sufficiently sensitive to study the effect of a mixture of BPDE and arsenic on the damage and repair of DNA in A549 cells.

Experiments using only one carcinogen may provide insight into the carcinogen's effect on human health, but most people are exposed to a mixture of various environmental components and some of these are carcinogens in low concentrations. For example, unrefined coal tar induces earlier tumors and produce more malignant tumors than those induced by pure PAHs at much higher concentrations than the PAHs found in coal tar, suggesting that synergistic interactions may promote carcinogenesis (Rubin 2001).

BPDE has been recognized to be a human carcinogen and much work has been done to demonstrate the effects of BPDE in cells *in vitro* and *in vivo*. (Galati et al. 2001; Li et al. 2001b; Venkatachalam, Denissenko, and Wani 1995; Maier et al. 2002; and many others). Investigations into mixtures such as arsenic and BPDE may provide insight into combined effects.

## MATERIALS AND METHODS

### Reagents

(±)-anti-r-7,t-8-dihydroxy-t-9,10-epoxy-7,8,9,10-tetrahydrobenzo[α]pyrene (BPDE) was provided by the National Cancer Institute Chemical Carcinogen Reference Standard Repository (Midwest Research Institute, Kansas City, Mo.). Benzo[α]pyrene was purchased from Sigma (Oakville, Ont.). Mouse monoclonal anti-BPDE-dG antibody (8E11) was purchased from BD Pharmingen (San Diego, Calif.). Secondary antibodies conjugated with fluorescent dye (goat anti-mouse antibody fragment Alexa Fluor 546) were purchased from Molecular Probes (Eugene, Ore.). Oligonucelotides were purchased from Integrated DNA Technologies (Coralville, Iowa). DNA was extracted with DNAzol reagent purchased from Life Technologies, GibcoBRL. Dulbecco's Modified Eagle's Medium/Ham's F12 (D-MEM/F12) cell media and Tris buffer were purchased from GibcoBRL. Arsenic standard was obtained from Aldrich (Milwaukee, Wis.). Solvents and other biochemicals were supplied by Sigma (St. Louis, Mo.) and Fisher Scientific (Fair Lawn, N.J.).

**Preparation of 190-mer Oligonucleotide Standard**

A 190-mer oligonucleotide containing a single BPDE adduct was synthesized using a similar approach to that described in Chapter 4. This is a longer oligonucleotide compared with the 90-mer used in Chapter 4. The main reason for using this longer oligonucleotide is that it is a more suitable standard for the non-competitive immunoassay of BPDE adduct in cellular DNA.

This standard oligodeoxynucleotide has 190 bases and is modified by covalently binding a single BPDE on $N^2$-dG in the middle of the chain. The 190-mer was generated by ligation of (+) *trans*-BPDE-$N^2$-dG-16-mer (oligo 2) and two 87-mers (oligos 1 and 3) with a complementary 30-mer (oligo 4) (Figure 5.1). The 16-mer containing a single (+) *trans*-$N^2$-dG-BPDE adduct (oligo 2) was the same as that used in Chapter 4.

The oligo 1, oligo 3 and oligo 4 of 400 pmole each were mixed with oligo 2 (BPDE-16-mer) of 240 pmole in a ligation buffer (80 μL). The solution was heated to 95 °C for 5 min and then allowed to cool to room temperature over 1 h to anneal the oligonucleotides. T4 DNA ligase (2.0 μL, 8.5 Weiss unit/μL) was added and the sample incubated overnight at 16 °C for ligation.

After ligation, phenol/chlorform/isoamyl alcohol (25:24:1, pH 6.7±0.2, Cat# BP1752[I]-100, Fisher Scientific) was added to denature the excess enzymes. The sample was centrifuged at 1000 rpm for 10 min. The supernatant containing the oligonucelotide was back extracted three times with equal volumes of 10 mM Tris-HCl buffer, pH 8.0. The extracted oligonucleotide solutions were pooled and oligonucleotides were precipitated by adding 0.1 volume of 3 M sodium acetate buffer (pH 5.2) and 3 volumes of ice-cold 95% ethanol (containing 10 mM $MgCl_2$). The sample was centrifuged for 10 min at 14000 rpm and 4 °C, and the pellets were collected and washed twice with 95% ethanol. The pellets were dried in air, and redissolved in 20 μL loading buffer for subsequent gel electrophoresis purification.

Purification of BPDE-oligonucleotide was achieved using 8% denatured (7 M urea) polyacrylamide gel electrophoresis (PAGE). Electrophoresis was carried out at 200 V (~19 mA) for 45 min on a 8 cm × 4 cm × 1 mm polyacrylamide gel. The running buffer was 1×TBE (89 mM Tris, 89 mM boric acid, pH 8.3). The gel slice containing the target BPDE-oligonucleotide was cut and the BPDE-oligonucleotide was recovered according to the published protocols (Chory and Pollard Jr. 1999). The BPDE-oligonucleotide was dissolved in ultrapure water and its concentration was determined by absorbance at 260 nm.

Oligo 1: 5' –CC CCC CCC CCC CCC CCC CCC CCC CCC CCC CCC CCC CCC
CCC CCC CCC CCC CCT TAA GCT TCC TCA ACC ACT TAC
CAT ACT CGA GAT T-3' (87 mer)
Oligo 2: 5'-CCC ATT ATG(BPDE) CAT AAC C-3' (BPDE-16-mer, 5' phosphorylated)
Oligo 3: 5' -/5phos/ GTC ATA TGC CGC CTC TGA CCT TCC TAG AAT TCC ATC
CCC CCC CCC CCC CCC CCC CCC CCC CCC CCC CCC CCC CCC
CCC CCC CCC CCC -3' (87 mer)
Oligo 4: 5'-CAT ATG ACG GTT ATG CAT AAT GGG AAT CTC (30-mer)

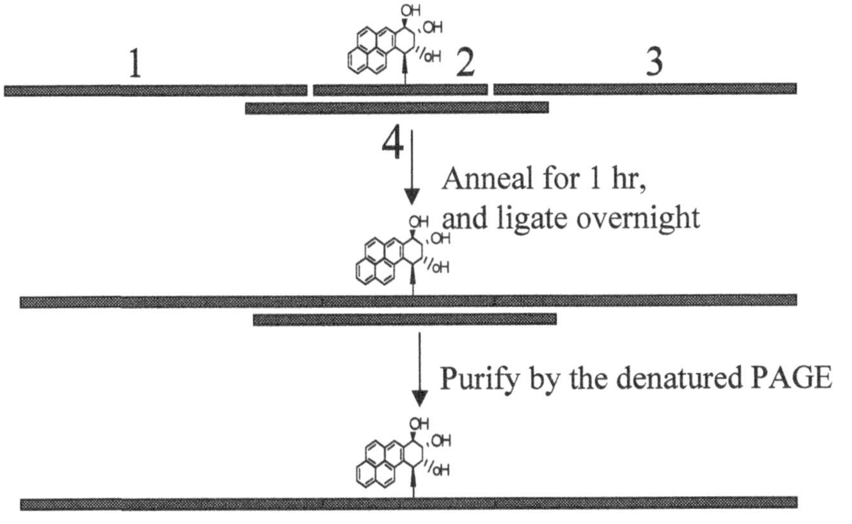


Figure 5.1. Schematic diagram showing the preparation of a 190-mer containing a single BPDE adduct. A 16-mer oligonucleotide (oligo 2) contained a single BPDE adduct. This oligonucleotide and two other 87-mer oligonucleotides (oligo 1 and oligo 3) were ligated with a complementary oligonucleotide (30-mer, oligo 4) to form a 190-mer oligonucleotide containing the single BPDE adduct.

## Instrumentation

A CE/LIF system similar to that illustrated in Chapter 4 (Figure 4.1) was used. Fused silica capillaries (Polymicro Technologies, Phoenix, Ariz.) with an internal diameter of 21 μm and an outer diameter of 150 μm were used for separation. The total capillary length was 25-36 cm. A green HeNe laser (543.5 nm) was used for fluorescence excitation. Fluorescence intensity at 580 nm was detected for quantitation.

## Treatment of A549 Cells with BPDE and Arsenite

The A549 cell line was maintained in DMEM/F12 medium (Gibco BRL, Gaithersburg, Md.) supplemented with 10% fetal bovine serum. The cells were seeded at $1 \times 10^5$ cells/plate and maintained at 95% humidity and 5% $CO_2$ for 20 h prior to the addition of BPDE or BaP. Old culture media were removed from each culture plate, and the cells were washed twice with phosphate-buffered saline (PBS). Media containing BPDE at various concentrations from 1 nM to 1 μM were added to the designated plates. The cells were further incubated in the media containing BPDE for 2 h. The cells were then washed with PBS prior to the addition of DNAzol lysis reagent (Gibco BRL) to facilitate cell lysis and DNA extraction. Subsequent steps involved a 99.9% ice-cold ethanol precipitation and a 70% cold ethanol wash to purify the genomic DNA. The final DNA concentration was measured at $OD_{260}$ using ultrapure $H_2O$ as a blank.

The concentrations of BaP tested were from 0.03 μM to 1 μM. Except that BaP was incubated with cells for 24 h, all other conditions were the same as those used for BPDE incubation with the cells.

For BPDE repair experiments, cells were incubated with BPDE for 4 h. The cells were then washed with PBS. Fresh media containing no BPDE was added. Following various periods of incubation, cells were lysed and DNA extracted for analysis.

To study the effect of arsenic on the induction and repair of BPDE-DNA adducts, A549 cells were first incubated in the D-MEM/F12 medium supplemented with arsenite for 24 h. The cells were then incubated in media containing both BPDE and arsenite for 4 h. The cells were either lysed immediately (to study the induction of BPDE damage) or incubated in BPDE-free media containing arsenite (to study the repair of BPDE). For repair experiments, cells were lysed and DNA extracted for analysis after various periods of incubation.

**Sample Preparation**

To prepare samples for analysis, 12.5 μL of DNA solution and 2.5 μL of 1x Tris/glycine buffer were pipetted into 0.2-mL tubes. The samples were placed in a heat block at 95°C for 5 min to denature the DNA, and then placed on ice for 10 min. Samples were removed from ice and incubated at room temperature with human IgG, the primary antibody and secondary antibody fragment in 1× Tris/glycine buffer (pH 7.5). The mixtures contained 80~700 μg/mL DNA, 2.0 μg/mL primary antibody, 2.0 μg/mL secondary antibody fragment, and 10.0 μg/mL human IgG. Addition of human IgG stabilized the antibody and enhanced the stability and formation of DNA-adduct complex (Wang et al. 2003). The samples were incubated at room temperature for 30 min, and subjected to CE/LIF analysis.

**Analysis with CE/LIF**

Samples were electrokinetically injected into the capillary by applying an injection voltage of 10 kV for 10 seconds. The separation was carried out at room temperature with a separation voltage of 15-20 kV. The running buffer was 1X Tris/glycine (30 mM Tris, 160 mM glycine), pH 8.5. After each analysis, the capillary was washed with 0.02 M NaOH for 7 min followed by water for 2 min and running buffer for 7 min by applying 15 kV. Capillary electrophoresis data were analyzed using IgorPro software (version 3.1, WaveMetrics Inc., Lake Oswego, Ore.).

## RESULTS AND DISCUSSION

### Capillary Electrophoresis Immunoassay for BPDE-DNA Adducts

The CE immunoassay mainly consists of two steps: immuno-reactions (DNA incubation with antibodies) and subsequent CE/LIF analysis of immuno-complexes. The two steps are schematically illustrated in Figure 5.2. During incubation (Figure 5.2a), the BPDE-DNA adduct is recognized by a specific primary (1°) antibody to the damage site of the DNA, which in turn is bound by a fluorescently labeled secondary (2°) antibody fragment at the Fc portion of the 1° antibody. Three fluorescent species can arise from the immuno-reactions: the 2° antibody alone; the complex of the 2° and 1° antibodies; and the complex of the 2° antibody, 1° antibody and DNA adduct (DNA adduct complex). In order to determine the complex of DNA adducts, the three fluorescent species are separated using capillary electrophoresis (Figure 5.2b). Subsequent laser induced fluorescence (LIF) detects the fluorescent species migrating out from the capillary.

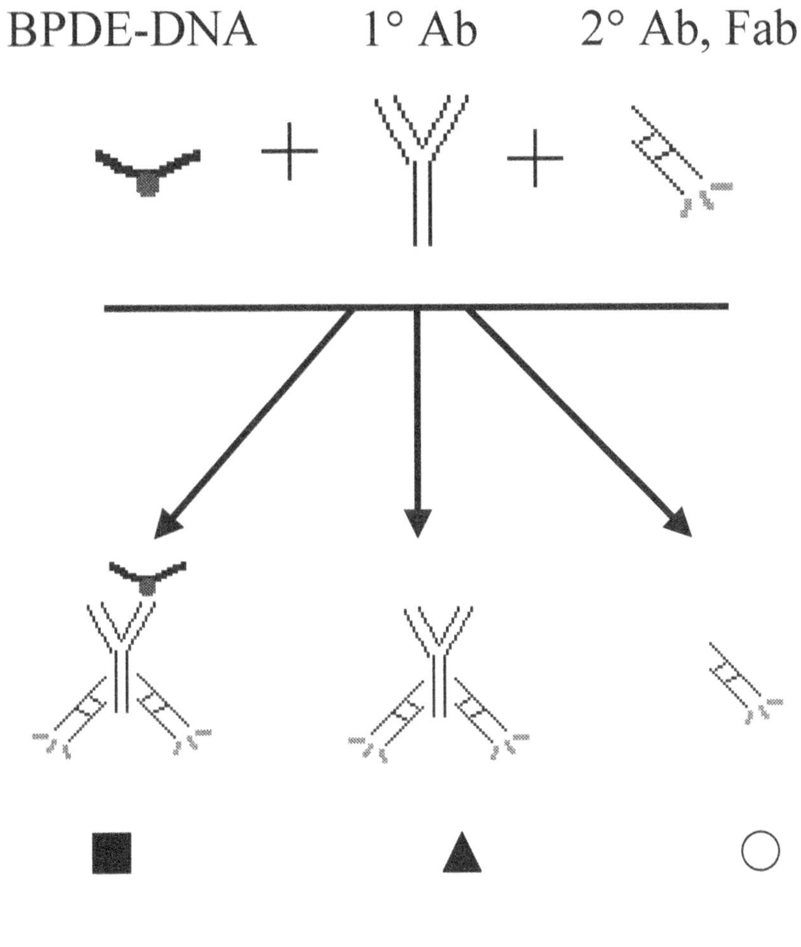


Figure 5.2a Illustration of immuno-reaction in the capillary electrophoresis immunoassay. After incubating DNA adducts with antibodies, three fluorescent species can be formed: 2° antibody (○); complex of 2° and 1° antibodies (▲); and complex of DNA adduct, 1° and 2° antibodies (■). The third one is the target immuno-complex with DNA adduct.

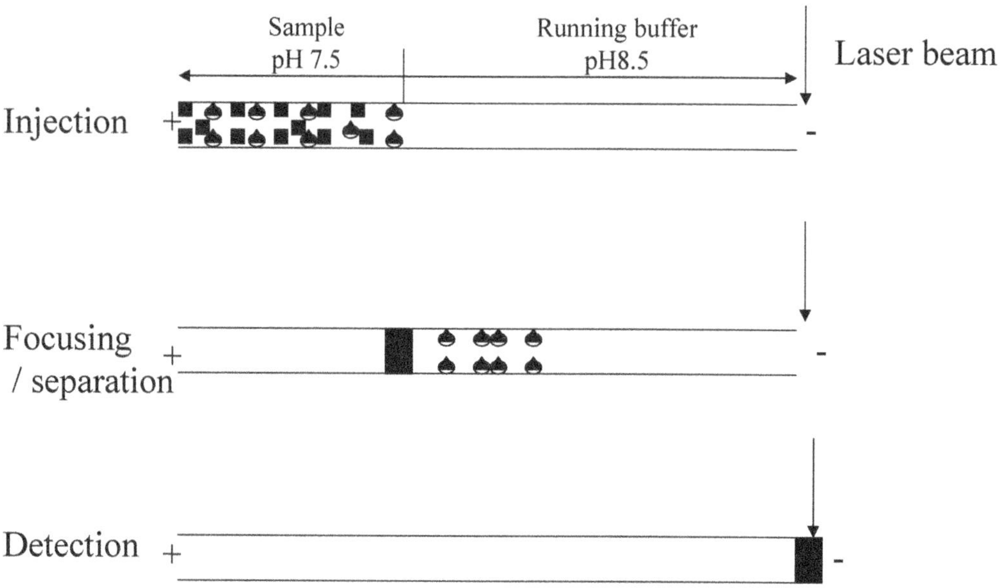


Figure 5.2b. Illustration of CE/LIF analysis of immuno-complex with DNA adduct in capillary electrophoresis immunoassay. The sample is injected into the capillary, and forms two pH zones: sample (pH 7.5) and running buffer (pH 8.5). The immuno-complex with DNA adduct is focused between the two pH zones and further separated by applying high voltage. The solutes are detected when they migrate out of the capillary.

Figure 5.3a shows a series of electropherograms from the CE/LIF immunoassay of a 190-mer oligonucelotide containing a single BPDE (BPDE-190mer). This oligonucleotide is a synthetic single stranded deoxynucleotide containing a single *trans*-BPDE-$N^2$-dG in the middle of the oligonucleotide. Mouse monoclonal anti-BPDE-dG antibody, 8E11 (MAb 8E11), recognizes the BPDE adduct and was chosen as the 1° antibody. A monovalent antibody fragment, Alexa Fluor 546 fragment of goat anti-mouse IgG (Fab), was used as the 2° antibody. This 2° antibody has better binding quality (monovalent binding site) and higher sensitivity (2-10 times) over other 2° antibodies.

Figure 5.3a shows that the complex (peak 2) of BPDE-190mer with the antibodies (1° and 2° antibodies) is well separated form the unbound antibodies (peak 1). While the free 2° antibody cannot be separated from its complex with 1° antibody (peak 1 at migration time 1.0 min), the binding with the DNA adducts changes the mobility of the complex (peak 2 at migration time 1.4 min). The peak shape for the DNA complex peak is much sharper than that of the antibodies. The separation efficiency of BPDE-190mer complex is over 1.6 million theoretical plates. The excellent efficiency is achieved with the DNA-induced focusing in capillary electrophoresis.

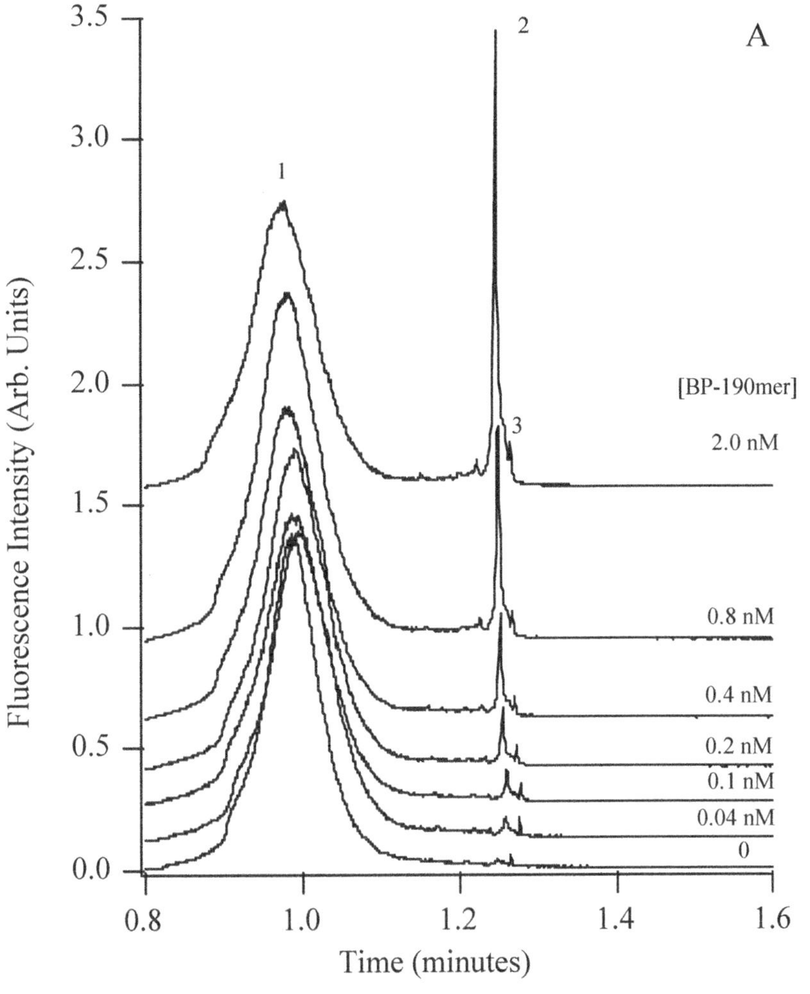


Figure 5.3a. A series of electropherograms from the CE/LIF analysis of the mixture of various concentrations of BPDE-190-mer incubated with 2° and 1° antibodies. The Alexa Fluor 546 Fab fragment of goat anti-mouse IgG was used as 2° antibody. The capillary was 29 cm long. Peak 1 is the mixture of 2° antibody and complexes of 2° and 1° antibodies, which overlap. Peaks 2 is the immuno-complex of BPDE-DNA adduct with 2° and 1° antibody. Peak 3 is an unknown impure fluorescent species present with 2° antibody.

Samples (pH 7.5) are injected into a capillary that is filled with slightly basic buffer at pH 8.5. There are two pH zones in the capillary, sample zone at pH 7.5 and the running buffer zone at pH 8.5. The complex of DNA adduct with the antibodies can be focused at the boundary between the two pH zones by DNA-induced focusing. This focusing technique requires the sample to have a lower pH than the running buffer. We have found that a sample pH of 7.5 and running buffer pH of 8.5 were optimum for the focusing. Incubation of the samples near physiological pH also enhances the formation of the stable DNA adduct complexes. As a consequence of the focusing of DNA adduct complex and the enhancement of the complex formation, the sensitivity of the CE immunoassay is greatly improved.

Figure 5.3 (b and c) shows the fluorescence response (in peak area and peak height) as a function of the concentration of the BPDE-190mer (0-2 nM). Both calibration curves are linear with regression coefficients of 0.9995. Based on a signal-to-noise ratio of 3, the limit of detection for BPDE-190-mer is estimated to be 5 pM. With an injection volume of 4 nL, the corresponding mass detection limit is approximately $2\times10^{-20}$ mole (~20 zeptomole). The improved sensitivity is very useful to accurately measure DNA damage at lower doses of exposure. The linear dynamic range can be extended to a concentration of 80 nM BPDE-190mer by using correspondingly higher concentrations of the 1° and 2° antibodies.

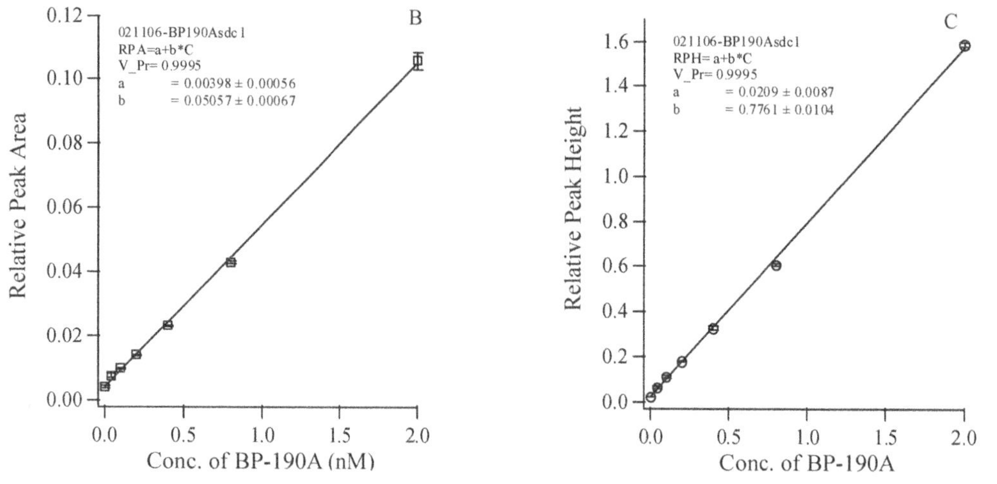


Figure 5.3 b and c. The relative peak area (a) and peak height (b) of the complex of BPDE-190-mer with 2° and 1° antibodies as a function of concentration of BPDE-190-mer. The data were collected from duplicate analyses.

### Assay for BPDE-DNA adducts in cellular DNA

The assay of BPDE adducts in cellular DNA is illustrated in Figure 5.3. The whole assay consists of three steps including cell exposure to carcinogen (BaP or BPDE), DNA extraction, and CE/LIF analysis. The human cells are first exposed to benzo[a]pyrene (BaP) or its reactive metabolite (BPDE). BaP is metabolized *in vivo* to the ultimate carcinogen, BPDE. BPDE reacts with DNA to form stable DNA adducts.

After exposure, the cells were harvested and treated with genomic DNA isolation reagent (DNAzol) to extract genomic DNA. The extracted DNA was denatured and then analyzed by CE immunoassays. The denaturation of genomic DNA to single strand DNA promotes the optimal binding of the DNA adducts with the specific antibody. This assay is specific for the DNA lesion of interest and it does not require DNA digestion or fluorescent labelling, making it an alternative to other more time-consumming and laborious assays.

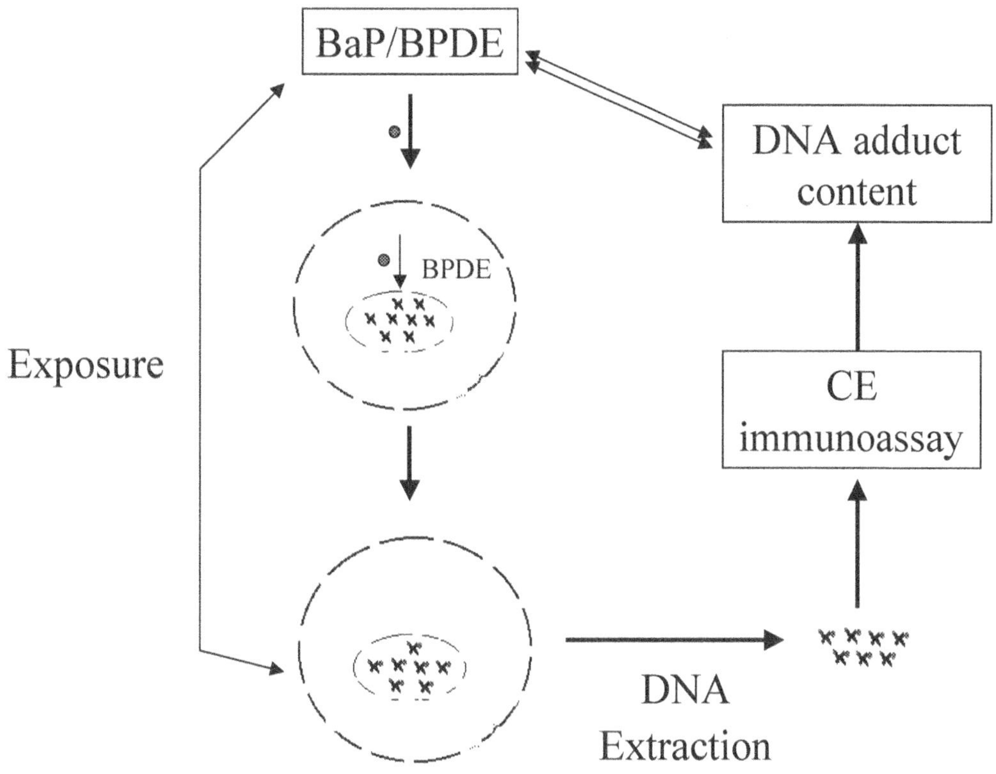


Figure 5.4. Schematic illustration of analysis of adducts in cellular DNA.

Figure 5.5 shows a series of electropherograms from the analysis of BPDE-DNA adducts in A549 cells that were exposed to BaP (0.03-1.0 μM ) for 24 h. The complex between DNA adduct and antibodies (peak 2) is well separated from the antibodies (peak 1). The separation efficiency for the DNA adduct complex (peak 2) is over 1 million theoretical plates per meter of capillary column. The excellent separation efficiency is due to a significant focusing of the DNA-antibody complex in the capillary during electrophoresis process. The 2° antibody and its complex with the 1° antibody are not resolved and appear overlapped (peak 1). This peak is broader, indicating that there is no significant focusing.

It is evident from Figure 5.5 that the signal of the DNA adducts (peak 2) increases with increasing concentration of BaP that was used to incubate with the A549 cells for 24 h. The signal from the unexposed DNA is very low and resulted from trace fluorescent impurity in fluorescently-labelled 2° antibody.

Using the calibration obtained from the BPDE-190mer standard under same conditions, the BPDE-DNA adduct signal obtained from the cells exposed to 0.03 μM BaP for 24 h corresponds to 2 adducts/$10^8$ nucleotides (nts). The detection limit, based on a signal-to-noise ratio of 3, is ~3 BPDE-DNA adducts per $10^9$ nts from the analysis using 2 ng DNA.

In our experiment, 1 million A549 cells were exposed to as low as 30 nM BaP in 3.0 mL medium. Under the given conditions, we detected 2 BPDE adducts/$10^8$ nucleotides. The absolute amount of BaP (23 ng) used in this experiment is comparable to that produced by the smoke of a single cigarette (20-40 ng) (Hecht 1999).

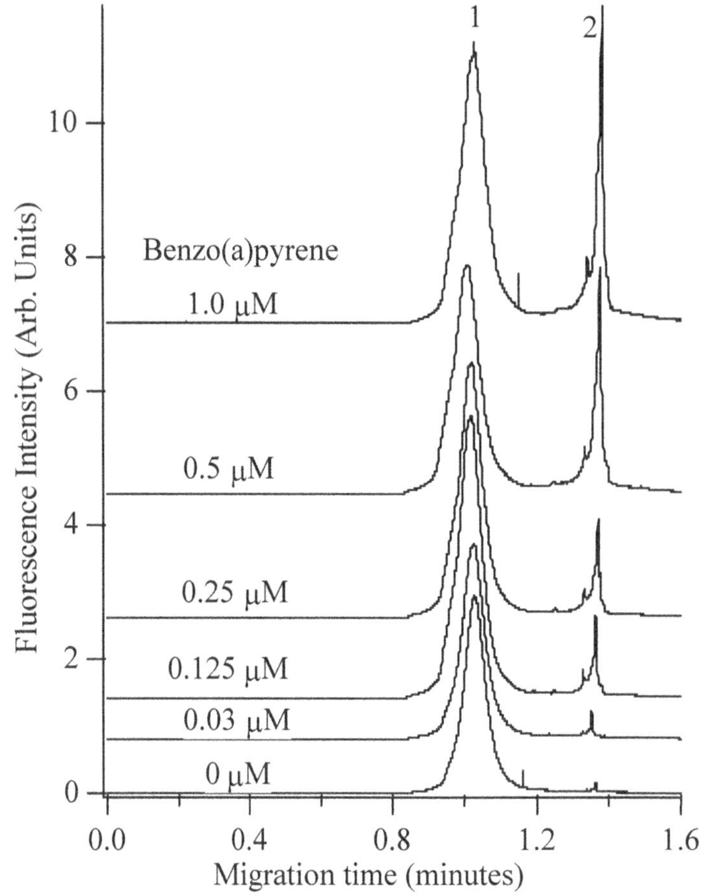


Figure 5.5. A series of electropherograms from CE/LIF analysis of BPDE-DNA adducts from A549 cells incubated with BaP from 0.03 μM to 1.0 μM for 24 h, showing the separation and focusing of DNA adduct complex with 2° and 1° antibodies. Samples were prepared in 1× Tris-glycine-HCl buffer at pH 7.5, and separated with buffer containing 30 mM Tris and 160 mM glycine at pH 8.5. The samples were electrokinetically injected into the uncoated fused-silica capillary (27 cm long, 20 μm i.d., and 150 μm o.d.) by applying 10 kV for 10 seconds and separated by applying 20 kV. Laser-induced fluorescence ($\lambda_{ex}$ = 543.5 nm and $\lambda_{em}$ = 580 nm) was detected from Alexa Fluor 546 labelled 2° antibody. Peak 1 is the mixture of 2° antibody and the complex of 2° and 1° antibodies, which overlap. Peak 2 is the immuno-complex of BPDE-DNA adduct with 2° and 1° antibody.

**Dose Responses of BPDE and BaP**

We measured the yield of BPDE-DNA adducts in human lung carcinoma (A549) cells that were incubated with either BaP or BPDE. We observed linear dose responses for both BaP and BPDE from 30 nM to 1 μM (Figure 5.6). A linear dose-response was previously observed for BPDE from ~50 nM to 1 μM by Wani et al. (2000) using an immunoslot blot assay. By comparing the slopes, the yield of adduct induced by BaP after 24 h exposure is 40-fold lower than that induced by BPDE after 2 h exposure. The result is consistent with previous work, which reported that BPDE was 40-50 times more active than BaP as an inducer of lung adenomas (Rubin 2002).

The lower level of BPDE-DNA adduct in A549 cells formed by BaP is most likely due to the metabolism steps required. While BPDE can directly react with DNA to produce BPDE-DNA adducts, BaP must first be metabolized through multi-stage enzymes to BPDE. It has been reported that <1% of BaP is metabolized to BPDE *in vivo* (Conney et al. 1994, Jyonouchi et al. 1999). This is consistent with the levels of BPDE-DNA adducts we observed in cells treated with BPDE and BaP.

We have detected ~24 BPDE-DNA adducts per $10^6$ nts in A549 cells that were incubated with 1 μM BPDE. This can be compared with the adduct levels measured by others as summarized in Table 5.1. Various cells were incubated with 1 μM BPDE for 30 min to 2 h and the adducts measured using various techniques. The measured adduct levels varied from 1.5 to 30 adducts per $10^6$ nucleotides. Among these results, the adduct levels measured by $^{32}$P-post-labelling are generally lower (8-40-fold) than those obtained by other methods. It has been reported that the adduct levels are underestimated by $^{32}$P-postlabelling assay (Pavanello et al. 1999, Phillips and Castegnaro 1993, Beland et al. 1999). For example, several methods, including $^{32}$P-postlabelling assay, were used in an interlaboratory comparison to assess the DNA adduct levels in mice that were treated with $^{3}$H radiolabeled BaP or 2-acetylaminofluorene (Phillips and Castegnaro 1993). For both carcinogens, the values determined by $^{32}$P-postlabelling assay were 4-8-fold lower than those determined by measuring $^{3}$H incorporation. Beland and coworkers found that the underestimation of adduct levels measured by $^{32}$P-postlabelling assay may result from the loss of the adducts upon chromatography and differences in hydrolysis and labelling efficiencies between the damage and normal nucleotides (Beland et al. 1999). Other variations in the adduct levels are probably due to the differences in cell type and treatment procedures.

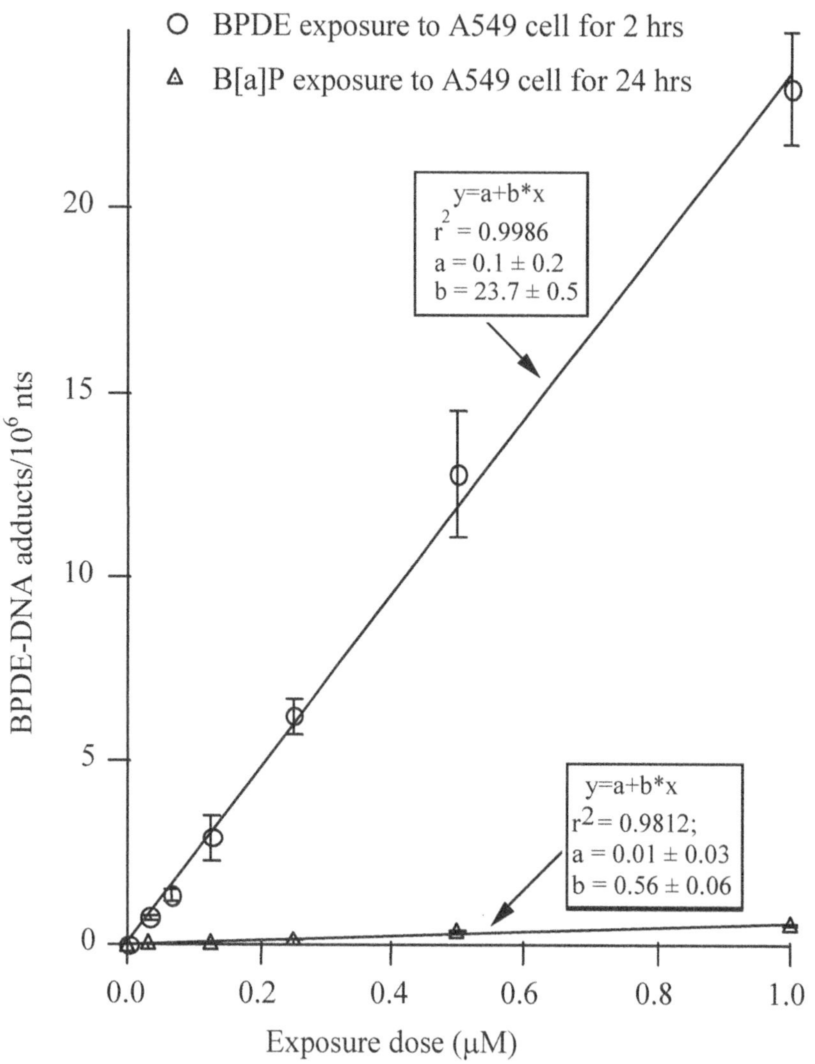


Figure 5.6. Dose-response curves from A549 cells incubated with BPDE for 2 h or BaP for 24 h. The data represent the mean ± standard deviation from three separate experiments.

Table 5.1
Summary of measured BPDE-DNA adduct levels from various experiments.

| Conc. of carcinogens | Exposed time (hour) | Adducts per $10^6$ nts | Assay method | Cell line | Culture/ incubation | Reference |
|---|---|---|---|---|---|---|
| 1.0 μM B[a]P | 24 | 18 | $^{32}$P-post-labeling/HPLC | Human MCF-7 | D-MEM/F12/ 10% fetal calf serum | * |
| 1.0 μM BPDE | 0.5 | 30 | Immunoslot blot assay | Normal human fibroblasts | Serum deficient medium, pH 7.0 | † |
| 1.2 μM BPDE | 1.0 | 1.5 | $^{32}$P-post-labelling/TLC | Human 041 TR | D-MEM/ serum free | ‡ |
| 1.0 μM BPDE | 0.5 | 17 | UvrABC nuclease assay | CHO AT3-2 | DPBS buffer | § |
| 1.0 μM BPDE | 2 | 2.5 | $^{32}$P-post-labelling/TLC | Human MRC5CV 1 | MEM/Earle's salts/10% fetal calf serum | ** |
| 1.0 μM BPDE | 2 | 24 | CE/LIF immunoassay | Human A549 | D-MEM/F12 medium/10% fetal calf serum | †† |

* Melendez-Colon et al. 2000
† Wani et al. 2000
‡ Lloyd and Hanawalt 2000
§ Tang, Pao, and Zhang 1994
** Hanelt et al. 1997
†† This work

**Removal of BPDE Adducts from the DNA of A549 Cells Exposed to 1 μM BPDE**

We further investigated the repair of DNA damage over time in the absence and presence of arsenic. Figure 5.7 shows peak areas for DNA adduct complex peaks obtained from the analysis of cellular DNA from A549 cells. The cells were exposed to 1 μM BPDE for 4 h followed by incubation in BPDE-free media for 0, 2, 4, 16, and 24 h. No arsenic was present in this set of experiments. DNA damage, demonstrated by peak area, appeared to be greatest at 4 h. The reduced levels of BPDE adducts at 16 and 24 h are probably due to the repair of DNA damage by the cells.

Kuljukka-Rabb et al. (2001) exposed MCF-7 cells to 2.5 μM BaP for up to 48 h and noticed an initial increase in adducts up to 12 h followed by a decrease in adducts after 12 h. Our results are consistent with their findings although different cell lines (MCF-7 vs. A549) and DNA damaging agents (BaP vs. BPDE) were used. Slower increase of BPDE-DNA adduct observed by Kuljukka-Rabb et al. (2001) is probably related to the metabolism step required for BaP.

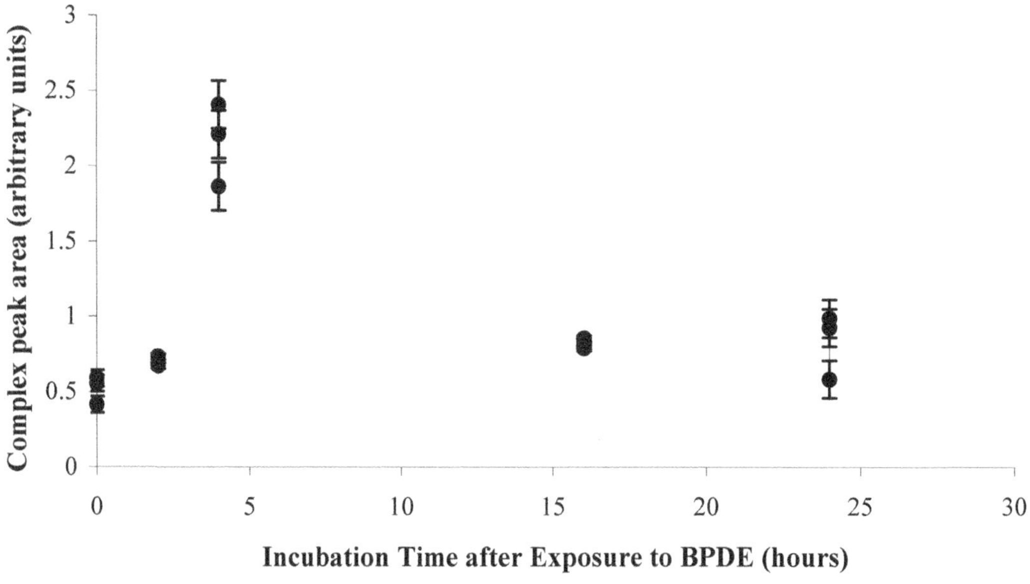


Figure 5.7. Peak areas of antibody-DNA adduct complex from the analysis of cellular DNA. A549 cells were exposed to 1 μM BPDE for 4 h and then incubated in BPDE-free media for 0-24 h. Each data point represents mean $\pm$ standard deviation from triplicate analyses of the samples.

**Removal of BPDE Adducts from the DNA of A549 Cells Exposed to 1 μM BPDE and 100 μg/L As$^{III}$**

Figure 5.8 shows the levels of BPDE adducts in A549 cells incubated with both BPDE and 100 μg/L (or 1.3 μM ) arsenite (As$^{III}$). The cells were pre-incubated for 24 h in the DMEM/F12 medium supplemented with 1.3 μM As$^{III}$. The medium was then changed to DMEM/F12 supplemented with 1 μM BPDE and 1.3 μM As$^{III}$ , and the cells were incubated for 4 h. Finally, the medium was changed back to DMEM/F12 medium supplemented with 1.3 μM As$^{III}$ , and the cells were incubated for 0-24 h.

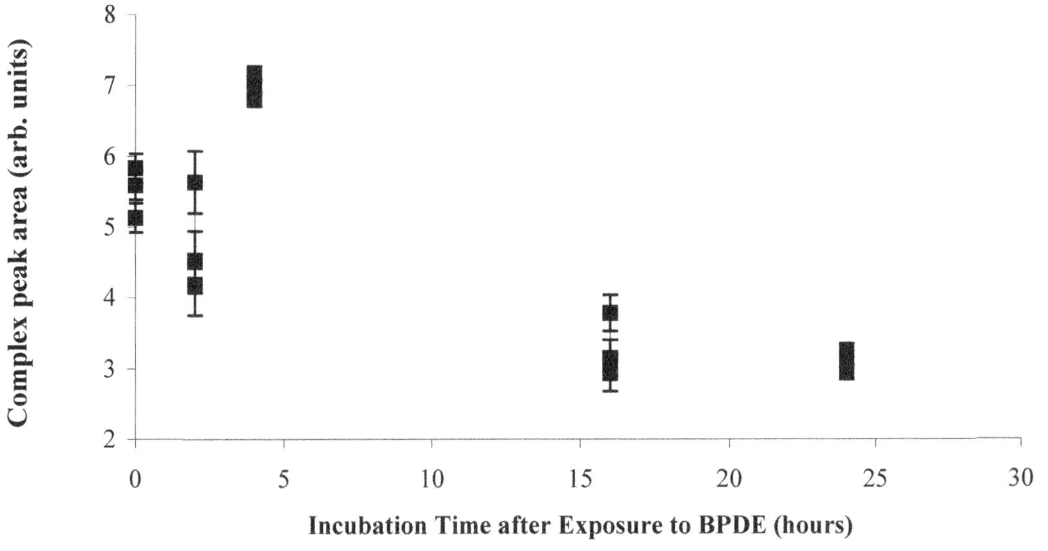


Figure 5.8. Peak areas of antibody-DNA adduct complex from the analysis of cellular DNA. The A549 cells were pre-incubated for 24 h in the DMEM/F12 medium supplemented with 1.3 μM $As^{III}$. The medium was then changed to DMEM/F12 supplemented with 1 μM BPDE and 1.3 μM $As^{III}$, and the cells were incubated for 4 h. Finally, the medium was changed back to DMEM/F12 medium supplemented with 1.3 μM $As^{III}$, and the cells were incubated for 0-24 h. Each data point represents mean ± standard deviation from triplicate analyses of the samples.

### Comparison of A549 Cells Exposed to 1 μM BPDE Alone and in Combination with 100 μg/L $As^{III}$

Table 5.2 summarizes results from the analysis of DNA adducts in A549 cells treated with BPDE alone or BPDE combined with $As^{III}$. At all time points BPDE adducts in cells treated with both BPDE and $As^{III}$ are greater than those from the cells treated with BPDE only. The greatest difference is at time 0, when the cells have had the least time to repair the damage. The combined treatment resulted in an approximately 10-fold increase in BPDE adduct levels compared with the cells treated with 1.0 μM BPDE only.

The effects of arsenic on BPDE-damaged DNA have recently been reported by others (Tran et al. 2002, Maier et al. 2002, Ho and Lee 2002). Like those of Maier et al. (2002), our experiments showed that cell treatment with BPDE (BaP in Maier's experiments) and arsenite resulted in increased DNA damage in comparison to cells treated with BPDE only. Our experiments used A549 human cells, while those of Maier et al. (2002) used Hepa-1 mouse cells. In both studies cells were pretreated with arsenite followed by simultaneous treatment with BPDE or BaP and arsenite.

Table 5.2
Comparison of peak areas of antibody-DNA adduct complex from the analysis of cellular DNA from A549 cells either treated with BPDE only or with a combination of BPDE and $As^{III}$.

| **Repair Time (h)** | 0 | | 2 | | 4 | | 16 | | 24 | |
|---|---|---|---|---|---|---|---|---|---|---|
| | **Complex Peak Areas** | | | | | | | | | |
| | BPDE + $As^{III}$ | BPDE | BPDE + $As^{III}$ | BPDE | BPDE + $As^{III}$ | BPDE | BPDE + $As^{III}$ | BPDE | BPDE + $As^{III}$ | BPDE |
| | 5.5908<br>5.1310<br>5.8298 | 0.4129<br>0.5901<br>0.5553 | 5.6350<br>4.5081<br>4.1906 | 0.7330<br>0.7426<br>0.5738 | 6.6840<br>7.1285<br>6.8121 | 2.8055<br>1.6626<br>2.407 | 2.9934<br>3.1484<br>3.6832 | 0.7894<br>0.8531<br>0.8055 | 2.1558<br>2.4359<br>2.5392 | 0.5846<br>0.9273<br>0.9869 |
| **Average** | 5.5 | 0.52 | 4.8 | 0.60 | 6.9 | 2.3 | 3.3 | 0.82 | 2.2 | 0.83 |
| **Ratio BPDE+ $As^{III}$ over BPDE** | 10 | | 8 | | 3 | | 4 | | 2.8 | |

Though DNA treated with arsenite and BPDE demonstrate greater complex peak areas than DNA from A549 cells treated with BPDE only, results in both cases indicate that adducts decrease with time (Figures 5.7 and 5.8). Likewise, Maier et al. (2002) found that while the presence of arsenite resulted in increased adducts, arsenite did not inhibit removal of BPDE-DNA adducts. As with our results, Maier et al. (2002) found that adduct levels in cells receiving both arsenite and BaP did not decrease to the level of adducts found in those cells treated with BPDE only during the time periods tested in these experiments.

In order to ensure that arsenic was present in the cell when BPDE was added, cells that would be exposed to BPDE and arsenite simultaneously were first pretreated with arsenite for 24 h. Hartwig et al. (1997) showed pretreatment to be a necessary step to ensure that arsenite will be taken up by the cells. Using human fibroblasts, Hartwig et al. (1997) exposed VH16, VH25, and XPC fibroblasts to a 10 μM concentration of arsenite for periods of up to 24 h and found that after 16 h the intracellular arsenite concentration did not increase (Hartwig et al. 1997). A 24 h pretreatment period was chosen also based on results from Chapter 2, which demonstrated a 90% cell survival rate for A549 cells exposed to 100 μg/L arsenite.

Some possible explanations for the differences between cells exposed to BPDE only and those exposed to both BPDE and arsenite follow. Hartwig et al. (1997) have suggested that arsenite interferes with the incision steps in nucleotide excision repair that allows the bulky adduct to be removed. The effect of arsenic then would be to decrease the rate of repair. In application to results seen in Figures 5.7 and 5.8, kinetics indicates that arsenic is not impeding repair. We propose that the higher level of damage in the arsenic treated cells may be due to chromatin changes that enhance BPDE reaction or alternatively arsenic reduces the levels of species such as glutathione that competes with DNA for reaction with BPDE. This will need to be studied more extensively.

In light of the varied responses of cell lines to DNA damaging agents, it is obvious that a generalization to the response of an organism to chemical insult can not be made easily. Factors such as genetic predisposition, race, and exposure frequency and duration can provide some guidelines but are not necessarily determinants of disease occurrence. Animal studies may possibly allow further insight into the effect of BPDE and arsenite on DNA. Tran et al. (2002) injected Sprague-Dewley rats with 10 mg/kg arsenite with or without 800 μg of BaP. They found that rats in both groups euthanized on day 5 (out of 1, 3, 5, 10, 27 days) showed maximum adduct formation. By day 10, BPDE adducts in the BaP only exposure group decreased approximately to initial levels while those rats exposed to both arsenite and BaP showed approximately the same number of adducts as seen on day 5. Interestingly, while the adducts in the combined treatment group (BaP + arsenite) persisted, they did not reach the higher adduct levels demonstrated in rats receiving BaP only. While *in vivo* results are applicable to the whole organism, these results do not necessarily indicate a similar human response.

An investigation into the metabolism of BPDE and arsenite in A549 cells may remove some uncertainty and provide a better understanding of the cellular response. In addition, further investigation into cytotoxicity, the exposure concentration-repair relationship, and cellular uptake and metabolism of BPDE and arsenic may allow further insight into the cellular response to damage.

Our research has opened up opportunities for further detailed studies of arsenic health effects. Additional work will include an investigation into the effects of different arsenic species on various cell lines. Experiments that investigate the effects of BPDE in combination with other trivalent arsenic species, such as $MMA^{III}$ and $DMA^{III}$, would be interesting in view of recent findings on arsenic metabolism. Comparison of DNA repair using both DNA repair deficient and proficient cells will provide useful information regarding the effect of arsenic on DNA repair process. Effect on cell cycle, replication, and metabolic products including reactive oxygen species will also be considered.

# CHAPTER 6
# ARSENIC SPECIATION ANALYSIS AND SPECIES STABILITY

## INTRODUCTION

Arsenic species ingested from drinking water are mainly inorganic arsenate ($As^V$) and arsenite ($As^{III}$). The inorganic arsenic species undergo metabolism in the body. Two major metabolites, dimethylarsinic acid ($DMA^V$) and monomethylarsonic acid ($MMA^V$), along with the inorganic arsenic species have been commonly observed in human urine (Buchet, Lauwerys, and Roels 1981; Buchet and Lauwerys 1994; Foà et al. 1984; Vahter 1994a; Vahter et al. 1995a; Hopenhayn-Rich et al. 1996b; Le and Ma 1998). Urinary excretion is the major pathway for the elimination of arsenic from the body (Crecelius 1977; Freeman et al. 1979; Pomroy et al. 1980; Tam et al. 1979; Vahter 1983; Buchet and Lauwerys 1994; Le, Cullen, and Reimer 1993, 1994; Yager, Hicks, and Fabianova 1997). Thus, speciation of arsenic in urine has been considered a measure of recent exposure to arsenic.

Speciation of arsenic has contributed much to the understanding of arsenic metabolism. It is now accepted that biomethylation is the major metabolic process for inorganic arsenic (Aposhian 1997; Cullen and Reimer 1989; Styblo, Delnomdedieu, and Thomas 1995; Vahter 1999; NRC 1999; Yamauchi and Fowler 1994; Goyer 1996). The stepwise methylation process is believed to involve a sequence of a two-electron reduction of arsenic followed by oxidative addition of a methyl group, as outlined previously in Figure 1.1 (Cullen, McBride, and Reglinski 1984; Cullen et al. 1989; Cullen and Reimer 1989). $TMAO^V$ and $TMA^{III}$ are the end product produced by some microorganisms. $DMA^V$ is the usual end product detected in humans, although $DMA^{III}$ was recently detected in human urine (Le et al. 2000a; Del Razo et al. 2001; Mandal, Ogra, and Suzuki 2001).

Most previous studies have focused on the speciation of relatively stable arsenic compounds that are readily detected in human urine (Buchet, Lauwerys, and Roels 1981; Crecelius 1977; Del Razo et al. 1997; Foà et al. 1984; Hakala and Pyy 1995; Hopenhayn-Rich et al. 1996b; Vahter et al. 1995a; Ng et al. 1998; Vahter 1999). However, the determination of the intermediate metabolites, monomethylarsonous acid ($MMA^{III}$) and dimethylarsinous acid ($DMA^{III}$), is also important to the understanding of the arsenic methylation pathway (Cullen and Reimer 1989, Zakharyan et al. 1999, Aposhian et al. 1999, Styblo et al. 1999, Sampayo-Reyes et al. 2000). Furthermore, recent studies have shown that the trivalent arsenic methylation metabolites are as toxic as or even more toxic than the inorganic arsenic species (Petrick et al. 2000; Styblo and Thomas 1995; Styblo, et al. 1997; Lin, Cullen, and Thomas 1999; Petrick et al. 2001; Mass et al. 2001). Thus, there has been much interest in the determination of these methylation intermediary metabolites in humans. The first component of this chapter describes our efforts in the speciation of arsenic metabolites including the intermediates in human urine.

Another important issue regarding the speciation of $MMA^{III}$ and $DMA^{III}$ is the stability and preservation of these species. There is no information on the oxidative stability of these species in samples during sample storage and handling. We report in this chapter on the stability of $MMA^{III}$ and $DMA^{III}$ in water and human urine samples that are stored at room temperature, 4 °C, and -20 °C for up to 5 months. We found that $MMA^{III}$ and $DMA^{III}$ were less stable than the other arsenic species, suggesting that new strategies for sample handling are needed for analysis of these trivalent arsenic species.

We further developed a technique that is able to preserve the speciation of $MMA^{III}$ and $DMA^{III}$ in human urine samples. To preserve the speciation of these two arsenicals, a metal complexing agent (diethylammonium diethyldithiocarbamate) was added to urine samples as a preservative, and the pH of the samples was adjusted to 6.0. Samples were stored at -20 °C and arsenic species were found to be stable for up to two months.

## INSTRUMENTATION

### HPLC separation with hydride generation atomic fluorescence detection (HPLC-HGAFS)

Separation of various arsenic species was carried out using a HPLC system that consisted of a HPLC pump (Model 307, Gilson, Middletone, Wis.), a 6-port sample injector with a 20-μL sample loop (Model 7725i, Rheodyne, Rohnet Park, Calif.), and an appropriate column (Le and Ma 1998). The column was mounted inside a column heater (Model CH-30, Eppendorf, Westbury, N.Y.) which was controlled by a temperature controller (Model TC-50, Eppendorf). The column temperature was maintained at $50\pm1$ °C. Mobile phase was pre-heated to the temperature of the column by using a precolumn coil of 50 cm stainless steel capillary tubing, which was also placed inside the column heater.

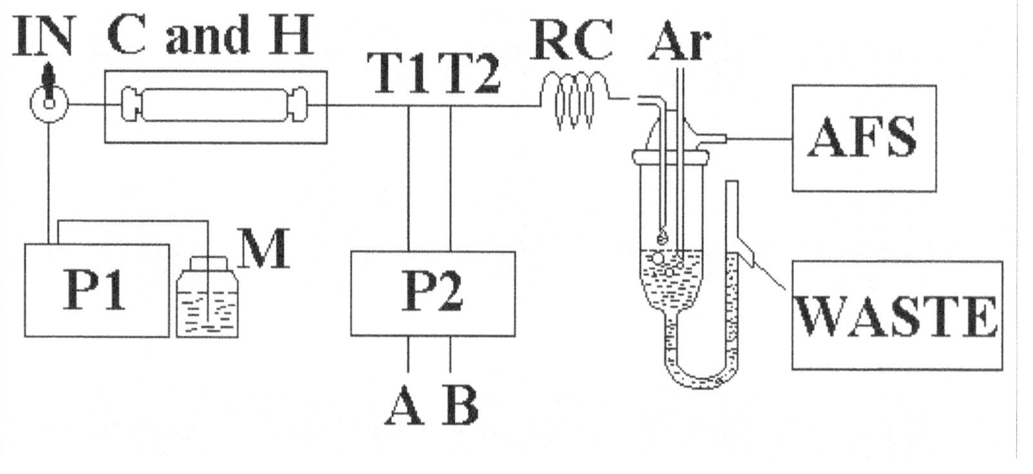


Figure 6.1. A schematic diagram showing the HPLC-HGAFS system for arsenic speciation analysis. (*Source:* Ma and Le 1998.)

*Notes*: A, acid (1.2 M HCl); AFS, atomic fluorescence detector; Ar, argon carrier gas; B, 1.3% $NaBH_4$ in 0.1 M NaOH; C and H, column and column heater; IN, sample injector; M, mobile phase; RC - reaction coil (50 cm long and 0.8 mm i.d.); T1 and T2, T-joints.

A hydride generation atomic fluorescence spectrometer (HGAFS) (Model Excalibur 10.003, P.S. Analytical, Kent, UK) was used for the detection of arsenic. The combination of HPLC and HGAFS is shown schematically in Figure 6.1 (Ma and Le 1998). Effluent from the HPLC column directly meets at two T-joints, with continuous flows of hydrochloric acid (1.2 M, 10 mL/min) and sodium borohydride (1.3%, 3 mL/min) introduced by using a peristaltic pump. Upon mixing, arsenic hydrides (arsines) generated from the reaction are separated from liquid waste in a gas/liquid separator apparatus and carried by a continuous flow of argon carrier gas (250 mL/min) to the atomic fluorescence detector. The atomic fluorescence detector consisted of a boosted discharge arsenic hollow cathode lamp as an excitation source, a hydrogen diffusion flame to atomize arsenic, and fluorescence detection optics. The hydrogen, produced as a hydride generation by-product, was sufficient to maintain the hydrogen diffusion flame that decomposes arsines and atomizes arsenic. Atomic fluorescence from arsenic was detected at a right angle using a solar blind photomultiplier tube. A Pentium computer with chromatography software (Star Workstation, Varian, Victoria, Australia) was used to record and process signals from the atomic fluorescence detector. IgorPro software (WaveMetrics, Lake Oswego, Ore.) was used to plot chromatograms.

**HPLC-Inductively Coupled Plasma Mass Spectrometry (HPLC-ICPMS)**

The HPLC separation system consisted of a Perkin Elmer 200 Series pump, a Perkin Elmer 200 Series autosampler, a Perkin Elmer 200 Series column oven and a reversed-phase C18 column (ODS-3, 150 mm x 4.6 mm, 3 μm particle size; Phenomenex, Torrance, Calif.). In addition, an ODS guard cartridge (4 mm long x 3.0 mm internal diameter, Phenomenex, Torrance, Calif.) was mounted before the analytical column to filter particles from the mobile phase and the samples to protect the analytical column. A solution (pH 5.85 or 6.5) containing 4.7 mM tetrabutylammonium hydroxide, 2 mM malonic acid and 4% (v/v) methanol was used as the HPLC mobile phase. The flow rate was 1.2 mL/min.

The column oven was used to control and maintain the HPLC column temperature. The mobile phase solution was preheated to the temperature of the column by a pre-column coil of stainless-steel capillary tubing, which was placed inside the column oven together with the separation column. The temperature of the column was maintained at 50 °C.

A Perkin-Elmer 6100DRC $^{plus}$ inductively coupled plasma-mass spectrometer (Perkin-Elmer/Sciex) was used as the HPLC detector. The standard liquid sample introduction system consisted of a Meinhard nebulizer coupled to a standard Cyclonic spray chamber. The HPLC column outlet was directly connected to the nebulizer using a short (20 cm) PEEK tubing (0.3 mm i.d.). Instrument settings were optimized for sensitivity with a solution containing 1 μg/L of each of Mg, In, Pb, Ce, Ba and Th. The general instrumental operating conditions are summarized in Table 6.1.

Turbochrom workstation software (Perkin-Elmer/Sciex) and IgorPro software (WaveMetrics) were used to convert ICPMS file and to plot HPLC data as chromatograms.

Table 6.1
ICPMS Operating Conditions

| | |
|---|---|
| RF power | 1100 W |
| Ar auxiliary gas flow rate | 1.2 L $min^{-1}$ |
| Ar plasma gas flow rate | 15 L $min^{-1}$ |
| Ar nebulizer gas flow rate | 0.90 L $min^{-1}$ |
| Dwell time | 200 ms |
| Detector mode | Dual mode |
| Scan mode | Peak hopping |
| Lens | Autolens On |
| Mass (m/z) | 75 |

*Source:* Jiang et al. 2003b

## STANDARDS, REAGENTS, AND SAMPLES

An atomic absorption arsenic standard solution containing 1000.0 mg As/L as arsenite (Sigma, St. Louis, Mo.) was used as the primary arsenic standard. Sodium arsenate, $As(O)OH(ONa)_2 \cdot 7H_2O$) and sodium cacodylate, $(CH_3)_2As(O)ONa$ were obtained from Sigma, and monomethylarsonate, $CH_3As(O)OHONa$, was obtained from Chem Service (West Chester, PA). The source of $MMA^{III}$ was the solid oxide ($CH_3AsO$), and $DMA^{III}$ was the iodide [$(CH_3)_2AsI$], which were prepared following literature procedures (Cullen et al. 1989, Burrows and Turner 1920). Trimethylarsine oxide [$(CH_3)_3AsO$, TMAO] was synthesized following the procedures of Merijanian and Zingaro (1966). Solutions of standard arsenic compounds were prepared by appropriate dilutions with deionized water from 1000 mg/L stock solutions. Solutions of $MMA^V$, $MMA^{III}$, $DMA^{III}$, $DMA^V$ and TMAO were standardized against a primary $As^{III}$ standard using an inductively coupled plasma mass spectrometer (Feldmann et al. 1999; Le and Ma 1998; Le et al. 2000a; Le, Ma, and Wong 1996).

Tetrabutylammonium hydroxide (TBAH), malonic acid, and disodium hydrogen phosphate ($Na_2HPO_4$) were obtained from Aldrich (Milwaukee, Wis.). HPLC grade methanol was from Fisher (Pittsburgh, Pa.). The HPLC mobile phase solutions were prepared in deionized water and filtered through a 0.45 µm membrane prior to use. Sodium borohydride (Aldrich) solutions (1.3%) in 0.1 M sodium hydroxide (Fisher) were prepared fresh daily. All reagents used were of analytical grade or better.

A Standard Reference Material (SRM), Toxic Metals in Freeze-Dried Urine SRM 2670, was obtained from National Institute of Standards and Technology (NIST, Gaithersburg, Md. ). The freeze-dried urine was reconstituted by the addition of 20.0 mL of deionized water as recommended by the supplier.

Urine samples were collected from 41 people in Inner Mongolia, China (Aposhian et al. 2000b) and 58 people in Romania (Aposhian et al. 2000a). Arsenic levels in their well water that was used for consumption were 510-660 µg/L for the Chinese group (Aposhian et al. 2000b) and 3-161 µg/L for the Romanian group (Aposhian et al. 2000a). They were asked to exclude seafood consumption for three days prior to and during the urine sample collection period. They stopped drinking well water and were provided with distilled water to drink. They were fasted overnight and then orally administered 300 mg of sodium 2,3-dimercapto-1-propane sulfonate (DMPS). DMPS is a chelating agent that is used to treat acute mercury and arsenic poisoning. A urine sample was collected from each participant before the administration of DMPS. Three urine samples were collected 0-2 h, 2-4 h, and 4-6 h after the administration of DMPS. Sample pH was adjusted to 4-5.

Urine samples were collected in 3-L polyethylene containers (Baxter Laboratories, Inc., Morton, Ill.), and the containers were then placed in portable iceboxes containing dry ice. The samples were kept frozen during transportation and were stored at -20 °C until just before the analysis when the samples were thawed at room temperature and an aliquot (20 μL) was analyzed for arsenic species using HPLC-HGAFS.

## PROCEDURES

### Speciation of $As^{III}$, $As^{V}$, $MMA^{V}$ and $DMA^{V}$ in Urine

Analyses of arsenic speciation in urine samples were carried out by using high performance liquid chromatography with hydride generation atomic fluorescence detection (HPLC-HGAFS). An aliquot of a sample was filtered through a 0.45 μm membrane prior to HPLC-HGAFS analysis. A reversed phase column (ODS-3, 150 x 4.6 mm, 3 μm particle size, Phenomenex, Torrance, Calif.) was used for separation. Mobile phase contained 5 mM tetrabutylammonium hydroxide, 4 mM malonic acid and 5% methanol (pH 5.8), and its flow rate was 1.5 mL/min.

### Speciation of $As^{III}$, $As^{V}$, $MMA^{V}$, $DMA^{V}$, $MMA^{III}$, and $DMA^{III}$ in Urine

A reversed phase column (ODS-3, 150 x 4.6 mm, 3 μm particle size, Phenomenex) was used for separation. A mobile phase solution (pH 5.9) contained 5 mM tetrabutylammonium hydroxide, 3 mM malonic acid, and 5% methanol, and its flow rate was 1.2 mL/min. The concentrations of hydrochloric acid (1.2 M) and sodium borohydride (1.3%) for hydride generation were optimized for maximum sensitivity of the less abundant arsenic species in urine, such as $MMA^{III}$, $DMA^{III}$, and $MMA^{V}$. Arsines generated were separated from liquid waste and carried by a continuous flow of argon to the atomic fluorescence detector for quantitation.

### Determination of TMAO

An anion exchange column (PRP X100, 140 x 4.1 mm, Hamilton, Reno, NV) was used to separate trimethylarsine oxide (TMAO) with a mobile phase containing 5 mM phosphate (pH 8.2) and 5% methanol. The flow rate of the mobile phase was 1.0 mL/min.

### Sample Storage Experiments

Measured amounts of $MMA^{III}$ and $DMA^{III}$ standards were dissolved in deionized water to prepare stock solutions (1000 mg/L of arsenic in the form of $MMA^{III}$ or $DMA^{III}$). Deionized water samples containing 100 μg/L of either $MMA^{III}$ or $DMA^{III}$ were prepared fresh by serial dilutions of the stock solutions with deionized water. Aliquots (1 mL) of these samples were placed in 1.5-mL plastic tubes, and separately stored at 25 °C (room temperature), 4 °C (refrigerated) and -20 °C (frozen), respectively.

A first-morning void urine was collected from a volunteer who had not eaten any seafood for two weeks prior to the sample collection. Urine samples from two other volunteers were also used for repeat experiments. Measured amounts of $MMA^{III}$ and $DMA^{III}$ stock solutions were spiked into the urine to make urine samples containing 100 μg/L of either $MMA^{III}$ or $DMA^{III}$. Aliquots (1 mL) of the urine samples were placed in 1.5-mL plastic tubes, and separately stored at 25 °C, 4 °C and -20 °C, respectively.

The deionized water and urine samples containing added $MMA^{III}$ and $DMA^{III}$ were stored for up to 5 months at the specified temperatures. They were periodically analyzed for arsenic speciation by using HPLC separation with hydride generation atomic fluorescence spectrometry detection (HPLC-HGAFS). Triplicate analyses were carried out from each sample aliquots.

## RESULTS AND DISCUSSION

### Speciation of the Usual Arsenic Compounds in Human Urine

We have previously investigated HPLC conditions and optimized them for rapid separation of $As^{III}$, $As^{V}$, $MMA^{V}$ and $DMA^{V}$ (Le and Ma 1998). The speciation of these four usual arsenic compounds in urine is complete in 4 min (Figure 6.2). The efficient separation within a short time was achieved by combining the use of a shorter HPLC column (15 cm) packed with smaller particles (3 μm) and the use of faster flow rate (1.5 mL/min).

The efficiency of a HPLC column is inversely proportional to the square of packing material diameter (Chen and Horvath 1995, Snyder and Kirkland 1977). Therefore, separation can be improved by using columns packed with smaller particles. Conventionally, 5-10 μm particle size is commonly used in HPLC columns. We used columns packed with 3 μm particles, and the improvement of separation efficiency enabled us to use shorter columns for the speciation of arsenic.

Furthermore, the optimum flow rate of the mobile phase is higher when a smaller size of particle is used in HPLC column packing. This is governed by the Van Deemter equation (Chen and Horvath 1995, Snyder and Kirkland 1977). A higher flow rate results in a faster HPLC separation.

Detection limits for arsenic species in urine matrix were 0.5 μg/L for $As^{III}$ and $MMA^{V}$ and 1 μg/L for $As^{V}$ and $DMA^{V}$. They were measured as arsenic concentrations corresponding to three times the signal-to-noise ratio. This detection capability allows for direct speciation of arsenic in human urine samples from the general population. Examples of the reported concentrations of arsenic in human urine from the general population are (mean $\pm$ standard deviation, μg/L) $9\pm7$ from a U.S. population (Kalman et al. 1990), $17\pm11$ and $11\pm6$ from European studies (Foà et al. 1984, Buchet et al. 1996), $21\pm7$ from Taiwan (Lin and Huang 1995), and $121\pm101$ from Japan (Yamauchi et al. 1989). People exposed to higher levels of arsenic from drinking water and food have corresponding higher levels of urinary arsenic, e.g. $56\pm13$ from Blackfoot disease patients (Lin and Huang 1995), $274\pm98$ from a highly exposed Argentina population (Vahter et al. 1995a), and 450-700 from a highly exposed Mexican group (Del Razo et al. 1997). Typical chromatograms from the analyses of urine samples are shown in Figures 6.2b and 6.2c. These samples were from two volunteers who did not have excess exposure to arsenic and who refrained from eating any seafood for three days prior to the collection of the first morning void. Concentrations of arsenic species in urine sample (b) were 4, 15, 4, and 2 μg/L for $As^{III}$, $DMA^{V}$, $MMA^{V}$, and $As^{V}$, respectively. Concentrations of arsenic species in urine sample (c) were 3, 23, and 3 μg/L for $As^{III}$, $DMA^{V}$, and $MMA^{V}$, respectively. $As^{V}$ in urine sample (c) was below detection limit.

A Standard Reference Material, SRM2670 (from NIST, Md.), was used for method validation and quality control purpose. This SRM consists of two components, normal level and elevated level of toxic metals. Figure 6.2d shows a chromatogram from the analysis of the SRM2670 urine containing normal levels of toxic metals. Results for the speciation of arsenic in this SRM using the HPLC-HGAFS method are $49 \pm 5$ for $DMA^{V}$ and $11 \pm 3$ for $MMA^{V}$. A reference value of total arsenic in this SRM is 60 μg/L. The other component of the SRM contains elevated levels of toxic metals and the certified value for total arsenic concentration is $480 \pm 100$ μg/L. Results for the speciation of arsenic using the HPLC-HGAFS method are $46 \pm 5$ for $DMA^{V}$, $11 \pm 3$ for $MMA^{V}$, and $460 \pm 25$ for $As^{V}$. These results are in good agreement with the certified and literature values (Crecelius and Yager 1997).

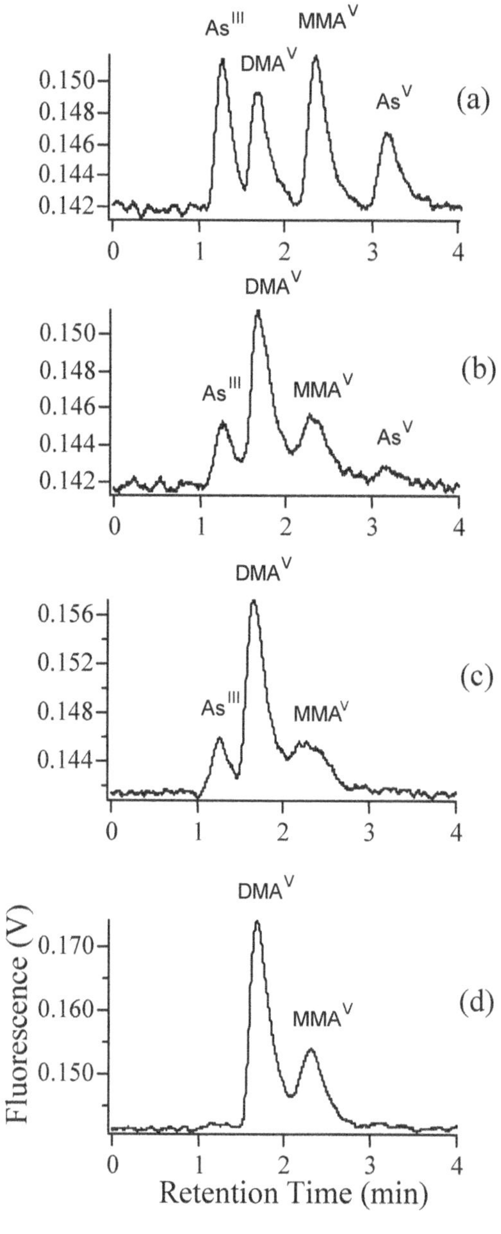


Figure 6.2. HPLC-HGAFS analyses of As$^{III}$, As$^{V}$, MMA$^{V}$, DMA$^{V}$ in a standard solution (a), a volunteer urine samples (b and c), and a standard reference material urine (d). (*Source:* Lu et al. 2001)

*Notes*: A reversed-phase column (Phenomenex ODS-3, 15 cm x 4.6 mm, 3 μm particle size) was used for ion paring separation. Mobile phase contained 5 mM tetrabutylammonium hydroxide, 4 mM malonic acid and 5% methanol (pH 5.8), and its flow rate was 1.5 mL/min. The HPLC column temperature was maintained at 50 $^{O}$C.

**Speciation of Intermediate Arsenic Metabolites in Human Urine**

The method described above is suitable for the rapid speciation of $As^{III}$, $As^{V}$, $MMA^{V}$ and $DMA^{V}$ (Le and Ma 1998) and has been demonstrated in pilot epidemiological studies of arsenic exposure and health effects (Calderon et al. 1999). We initially attempted to adopt this method for the speciation of $MMA^{III}$ and $DMA^{III}$ in human urine. However, $MMA^{III}$ coeluted with $As^{III}$ and $DMA^{V}$, and $DMA^{III}$ overlapped with $As^{V}$ within the narrow separation time window (4 min) (Le et al. 2000b). To obtain a separation of $MMA^{III}$ and $DMA^{III}$, we modified HPLC separation conditions to allow for a wider separation window. This was achieved by adjusting malonic acid concentration in the HPLC mobile phase. By reducing malonic acid concentration to 3 mM and a slight adjustment of the pH and flow rate of the mobile phase, an extended separation time period was obtained (Figure 6.3). The modified chromatography allowed for the separation of $MMA^{III}$ from $As^{III}$ and $DMA^{V}$, and the separation between $DMA^{III}$ and $As^{V}$ (Figure 6.3). This is at the expense of a longer retention time for the arsenic species; the speciation of the six arsenic compounds, $As^{III}$, $As^{V}$, $MMA^{V}$, $DMA^{V}$, $MMA^{III}$, and $DMA^{III}$ is complete in 6 min (Figure 6.3).

Figure 6.3 also shows chromatograms obtained from the analyses of a urine sample collected 4 h after a person ingested 300 milligram DMPS (dotted traces). Co-injection of the urine sample with authentic $MMA^{III}$ standard demonstrates the co-elution of the suspected $MMA^{III}$ in the sample with that of the standard $MMA^{III}$ (Figure 6.3b), confirming the identity of $MMA^{III}$ in the urine sample. Similarly, co-injection of the urine sample with standard $DMA^{III}$ (Fig. 6.3c) confirms the presence of $DMA^{III}$ in the sample.

We detected $MMA^{III}$ in 207 samples and $DMA^{III}$ in 9 samples out of 454 samples from 99 subjects. The highest $MMA^{III}$ concentration in the urine samples was 240 µg/L. In most cases, $MMA^{III}$ and $DMA^{III}$ were found in urine samples collected after the administration of DMPS. However, 10 samples collected before the administration of DMPS also contained detectable $MMA^{III}$ (Aposhian et al. 2000a). Figure 6.4 shows chromatograms from the speciation analysis of such a urine sample (Figure 6.4a) and the sample spiked with $MMA^{III}$ (Figure 6.4b). The urine sample was collected before the administration of DMPS. $MMA^{III}$ is clearly present in the urine sample from the subject who was not administered DMPS. Two samples collected before the administration of DMPS also contained detectable $DMA^{III}$.

Furthermore, analyses of some urine samples containing added DMPS (438 µM), $MMA^{V}$ (0.3 µM), and $DMA^{V}$ (0.7 µM) did not show evidence of $MMA^{III}$ and $DMA^{III}$ formation (Le et al. 2000a). These results demonstrate that the $MMA^{III}$ and $DMA^{III}$ detected in human urine samples are not due to endogenous reduction of $MMA^{V}$ and $DMA^{V}$ by DMPS.

The most commonly used hydride generation methodology does not distinguish between pentavalent and trivalent arsenic species unless carried out at selective pH conditions. The pentavalent arsenic species ($As^{V}$, $MMA^{V}$, $DMA^{V}$, and $TMAO^{V}$) require a low pH (usually pH $<$ 1) to form hydride upon treatment with sodium borohydride. At a higher pH ($>$5), only trivalent arsenic species ($As^{III}$, $MMA^{III}$, and $DMA^{III}$) form volatile hydrides. Thus, two separate analyses of each sample, carried out at different pH conditions, are required to selectively determine the trivalent arsenic species at the high pH and the total arsenic concentration at the low pH.

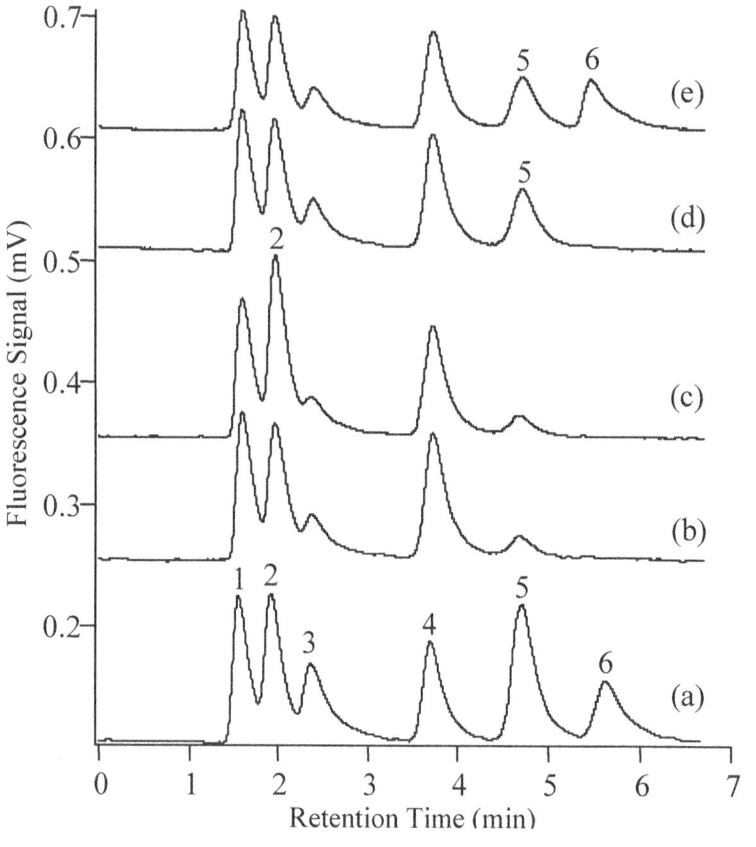


Figure 6.3. Typical chromatograms showing HPLC-HGAFS analyses of $As^{III}$, $As^{V}$, $MMA^{V}$, $DMA^{V}$, $MMA^{III}$, and $DMA^{III}$ in deionized water (a), a urine sample (b), and the urine sample spiked with $MMA^{III}$ (c), $DMA^{III}$ (d), and $As^{V}$ (e). (*Source:* Le et al. 2000a)

*Notes:* Separation was carried out on an ODS-3 column (15 cm x 4.6 mm, 3 μm particle size; Phenomenex) with a mobile phase (pH 5.95) containing 5 mM tetrabutylammonium hydroxide, 3 mM malonic acid, and 5% methanol. The flow rate of the mobile phase was 1.2 mL/min. The column was maintained at 50 °C. A hydride generation atomic fluorescence detector was used for detection of arsenic. Peaks labeled 1-6 correspond to $As^{III}$, $MMA^{III}$, $DMA^{V}$, $MMA^{V}$, $DMA^{III}$, and $As^{V}$, respectively. The urine sample was collected from a person 4 h after the administration of 300 mg sodium 2,3-dimercapto-1-propane sulfonate (DMPS). For clarity, chromatograms were manually shifted on vertical axis.

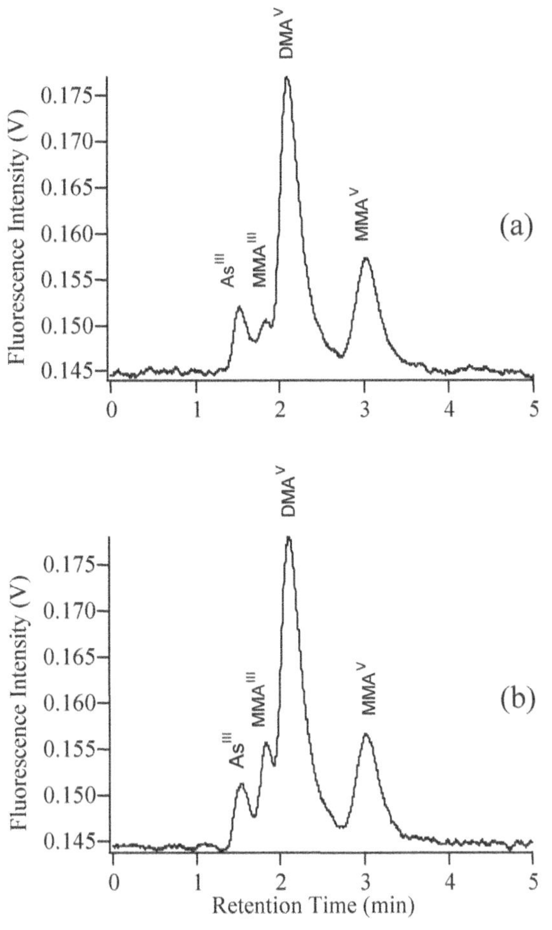


Figure 6.4. Chromatograms showing speciation analyses of arsenic in a urine sample (a), and the urine sample with addition of MMA$^{III}$ (b). (*Source:* Lu et al. 2001)

*Notes:* The urine sample was collected from a person who was not administered with DMPS. Separation was carried out on an ODS-3 column (15 cm x 4.6 mm, 3 μm particle size; Phenomenex) with a mobile phase (pH 5.95) containing 5 mM tetrabutylammonium hydroxide, 3 mM malonic acid, and 5% methanol. The flow rate of the mobile phase was 1.2 mL/min. The column was maintained at 50 °C. A hydride generation atomic fluorescence detector was used for detection of arsenic. Peaks labeled 1-6 correspond to As$^{III}$, MMA$^{III}$, DMA$^V$, MMA$^V$, DMA$^{III}$, and As$^V$, respectively. For clarity, chromatograms were manually shifted on vertical axis.

**Determination of TMAO**

Trimethylarsine oxide (TMAO) is an expected product of methylation of DMA$^{III}$, as shown in Scheme 1. We probed for the presence of TMAO to examine whether DMA$^{III}$ is further methylated to TMAO in humans. Under the ion pair chromatographic conditions described above, TMAO coeluted with As$^{III}$ and thus, could not be differentiated. Thus, we used a different separation mode, strong anion exchange chromatography (Figure 6.5). TMAO is resolved from the other arsenicals. Using this method, we analyzed all the urine samples that contained DMA$^{III}$, and we did not find detectable TMAO in any of the urine samples.

Figure 6.5 shows chromatograms from the speciation analysis of arsenic standards and a urine sample using strong anion exchange separation. The urine sample was obtained from a volunteer 4 h after the administration of DMPS and it contains both MMA$^{III}$ and DMA$^{III}$ as shown in Figure 6.3. Reanalysis of the same sample under the conditions suitable for TMAO analysis (Figure 6.5a) confirms that there was no detectable TMAO in the urine sample (Figure 6.5b).

TMAO has been identified as a metabolite in bacterial systems. Our recent speciation analyses of rat urine also identified TMAO, consistent with previous findings (Wanabuchi et al. 1996). However, little is known about TMAO in humans. Failure to observe TMAO in human urine suggests that either its concentration in urine is below the detection limit of the method or it is further metabolized to trimethylarsine (TMA$^{III}$) and subsequently exhaled in the breath, as it is a volatile species.

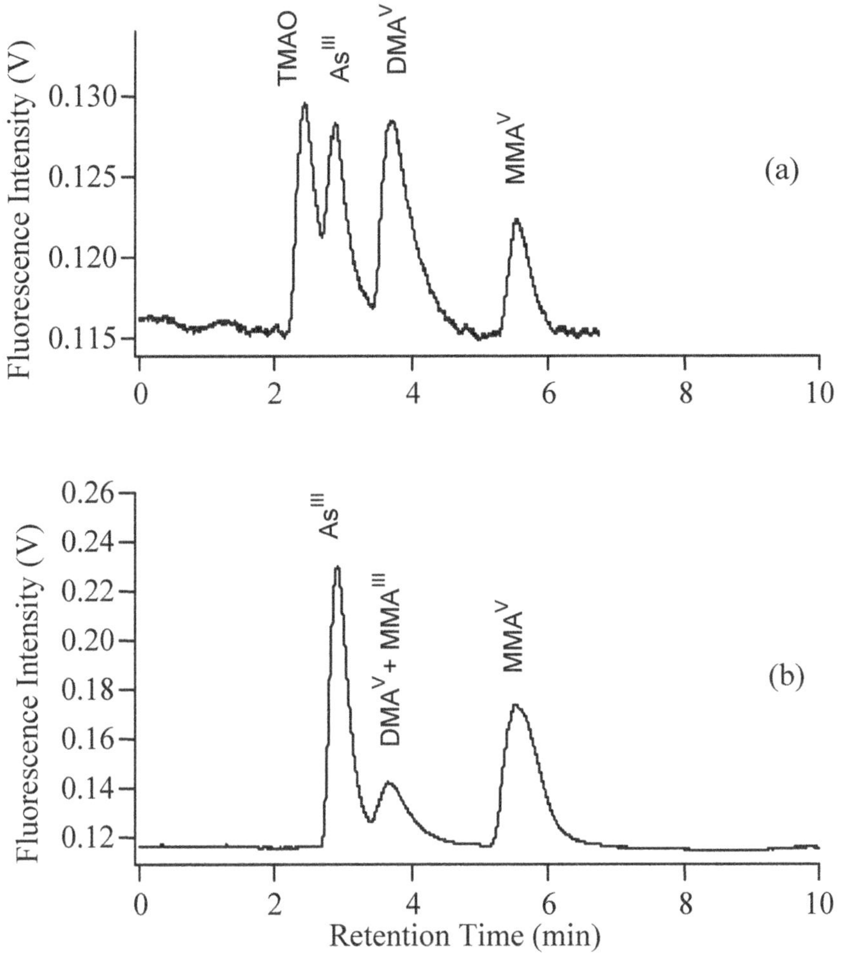


Figure 6.5. Chromatograms showing HPLC-HGAFS analyses of trimethylarsine oxide (TMAO) and a urine sample collected 4 h after the administration of DMPS. (*Source:* Le et al. 2000a)

*Notes:* Separation was carried out on a strong anion exchange column (PRP X-100, 15 cm x 4.1 mm) with a mobile phase containing 5 mM phosphate (pH 8.2) and 5% methanol. Note that TMAO was not detectable in the urine sample.

## Stability of $MMA^{III}$ and $DMA^{III}$

Both $MMA^{III}$ and $DMA^{III}$ were unstable in urine, although low temperature conditions (4 and -20 °C) improved the stability of these arsenic species over the room temperature storage condition (Gong et al. 2001). Table 6.2 shows the residual amounts of $MMA^{III}$ in water and urine samples after the samples were stored for one month and for 4 months. Deionized water and urine samples were spiked with $MMA^{III}$ (100 μg/L) and the samples were stored at -20, 4, or 25 °C. There was no $DMA^{III}$ detected in either deionized water or urine samples under the same storage conditions.

It is conceivable that $MMA^{III}$ and $DMA^{III}$ species are partially oxidized to $MMA^{V}$ and $DMA^{V}$ during sample collection, handling, and storage. The $MMA^{III}$ and $DMA^{III}$ detected in the urine samples may represent the residual amounts of these trivalent arsenic species. Therefore, the stability of $MMA^{III}$ and $DMA^{III}$ species under various storage conditions was studied in detail, as described below.

Table 6.2
Residual amounts of $MMA^{III}$ (μg/L) remained in the samples after each sample was spiked with 100 μg/L $MMA^{III}$ and the samples were stored for up to 4 months.

| Storage temperature (°C) | $MMA^{III}$ in water | | $MMA^{III}$ in urine | |
|---|---|---|---|---|
| | 1 month | 4 months | 1 month | 4 months |
| -20 °C | 90 ± 5 | 90 ± 5 | 38 ± 3 | 4 ± 2 |
| 4 °C | 100 ± 5 | 93 ± 5 | 5 ± 2 | n.d. * |
| 25 °C | 5 ± 2 | 5 ± 1 | n.d. * | n.d. * |

*Source:* Lu et al. 2001

* n.d.: Below detection limit of 2 μg/L.

## Stability of $MMA^{III}$ in Deionized Water

Figure 6.6 shows the changes of arsenic speciation over time when $MMA^{III}$ in deionized water was stored for up to 114 days. The pH of the water sample was 6.0. Approximately 15% of $MMA^{III}$ was oxidized to $MMA^{V}$ after storage at room temperature (25 °C) for only three days (Fig. 6.6a). and conversion from $MMA^{III}$ to $MMA^{V}$ continues to increase over time to approximately 80%. After 17 days, $As^{III}$ was also detected in the samples, which accounted for approximately 4-10% of the total arsenic. $MMA^{III}$ remained relatively constant at around 8% after 20 days.

$MMA^{III}$ is more stable when stored at -20 °C and 4 °C than at 25 °C. $MMA^{III}$ in deionized water can be stable for up to one month when stored at 4 °C in a refrigerator, after which time approximately 10% was oxidized to $MMA^{V}$ (Fig. 6.6b). Storing the sample at -20 °C did not further improve the stability of $MMA^{III}$ (Fig. 6.6c).

There seemed to be an equilibrium between $MMA^{III}$ and the oxidation product $MMA^{V}$ when the deionized water samples were stored at either -20 °C or 4 °C. After the relative concentration of $MMA^{V}$ reached approximately 10%, it remained relatively constant at this level. Approximately 90% $MMA^{III}$ remained in the solution and 10% was oxidized to $MMA^{V}$ after 4 months of storage at –20 °C and 4 °C (Fig. 6.6b and 6.6c).

Figure 6.7 shows representative chromatograms from the analysis of water samples (spiked with 100 µg/L $MMA^{III}$) stored for 5 months. Clearly, $MMA^{III}$ in the original sample was mainly oxidized to $MMA^{V}$ after storage at room temperature for 5 months (Fig. 6.7a). When the water sample was stored at 4 °C (Fig. 6.7b) and -20 °C (Fig. 6.7c), approximately 90% of $MMA^{III}$ remained in the sample and 10% was oxidized to $MMA^{V}$.

These results suggest that standard solutions of $MMA^{III}$ and $DMA^{III}$ that are commonly prepared in deionized water may be kept in a refrigerator or a freezer for at least one month without significant change in their speciation.

### Stability of $MMA^{III}$ in urine

Figure 6.8 shows the changes of $MMA^{III}$ in urine samples that were stored at 25 °C, 4 °C or –20 °C for up to 114 days. The urine was obtained from a volunteer who had not eaten any seafood for at least two weeks before collection of the first morning void. The pH of the urine sample was 5.5. The sample was supplemented with 100 µg/L of $MMA^{III}$, freshly prepared from the $MMA^{III}$ stock solution. The samples stored at room temperature showed complete conversion of $MMA^{III}$ to $MMA^{V}$ after 3 days (Fig. 6.8a). When $MMA^{III}$-spiked urine was stored at room temperature for 4 months, small amounts (<4%) of $As^{III}$ and $DMA^{V}$ were also detected along with $MMA^{V}$ (>96%) as the predominant species.

Storage at lower temperatures (-20 °C and 4 °C) slightly improved the stability of $MMA^{III}$ in urine. Approximately 30% of $MMA^{III}$ was converted to $MMA^{V}$ after one day of storage at either -20 °C or 4 °C. Further conversion of $MMA^{III}$ to $MMA^{V}$ in the urine samples stored at 4 °C was observed during the following 30 days at which time 98% was oxidized to $MMA^{V}$. Approximately 2% $MMA^{III}$ remained in the urine samples stored at 4 °C for up to 114 days (Figure 6.8b). The conversion of $MMA^{III}$ to $MMA^{V}$ in the samples stored at –20 °C was slower as shown in Figure 6.8c. Approximately 8% of $MMA^{III}$ (8 µg/L) remained in the samples after storage for 114 days.

Figure 6.9 shows typical chromatograms from the HPLC/HGAFS analyses of urine samples, stored for 5 months, at respectively 25, 4 or -20 °C. Although the urine samples were spiked with $MMA^{III}$ (100 µg/L), the major arsenic species after 5 months of storage was $MMA^{V}$. A small amount (~8%) of $MMA^{III}$ was clearly present in the urine samples stored at -20 °C.

The peaks in Figure 6.9 are broader than those shown in Figure 6.7. We found that this peak broadening was dependent on urine matrix. Our subsequent analyses of urine samples from two other volunteers showed that there was no peak broadening. Our previous study also showed that peak widths for arsenic species in deionized water and in most human urine samples were the same (Le et al. 2000a, 2000b). In the present study, the oxidative conversion behavior of $MMA^{III}$ to $MMA^{V}$ in all three urine samples was the same although HPLC peak widths were different.

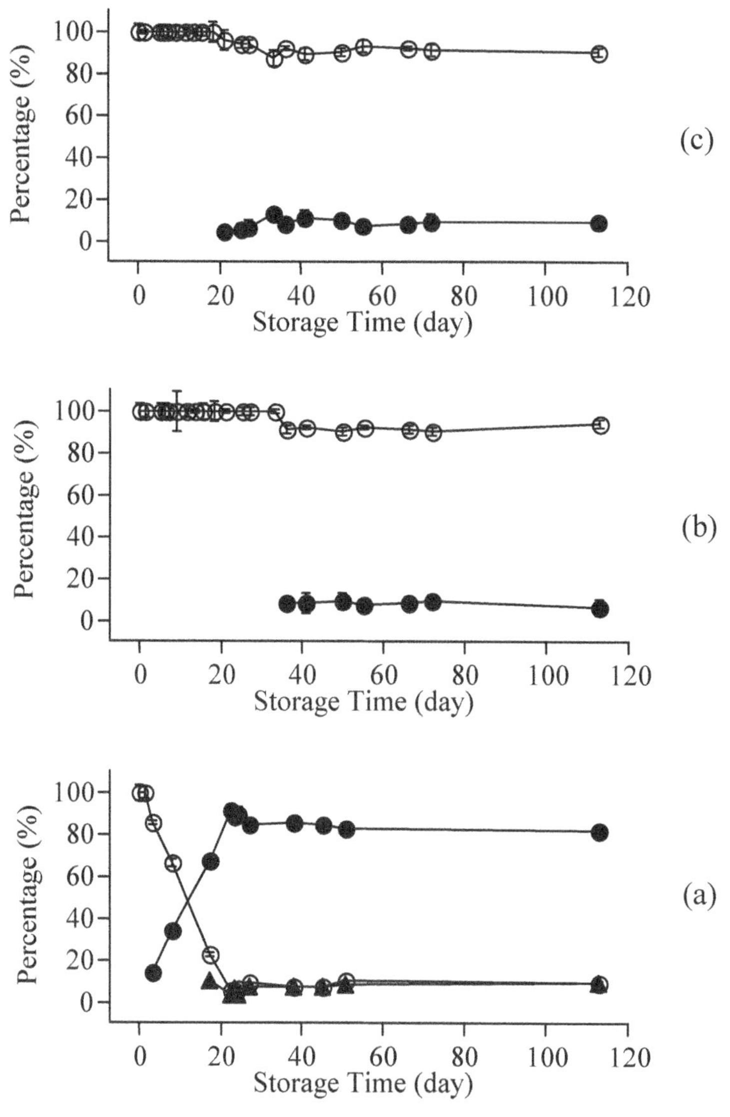


Figure 6.6. Effect of storage duration and temperature on the stability of $MMA^{III}$ in deionized water. (*Source:* Gong et al. 2001)

*Notes:* (a), 25 °C; (b), 4 °C; and (c), -20 °C.
(**O**), $MMA^{III}$; (●), $MMA^{V}$; (▲), $As^{III}$; (▼)

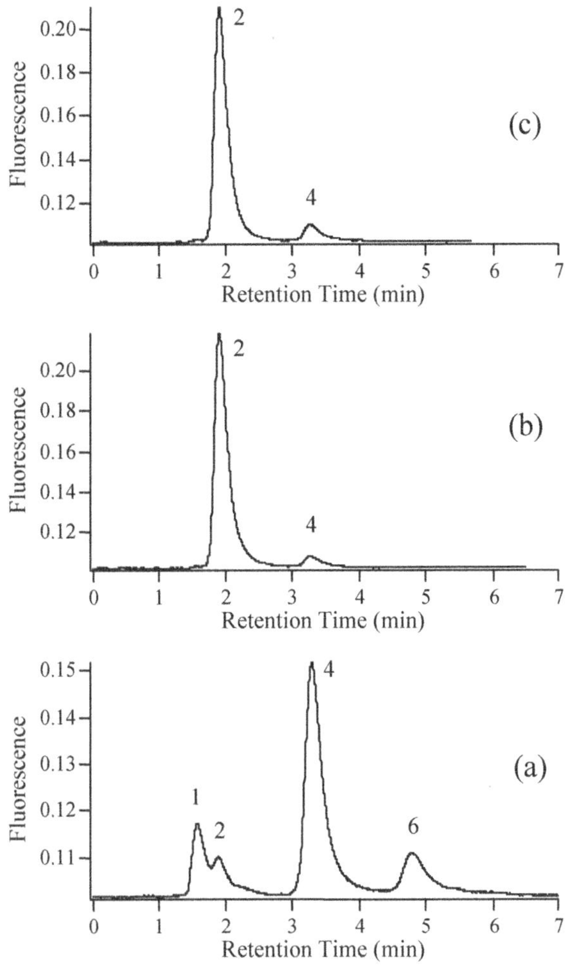


Figure 6.7. Typical chromatograms from the HPLC-HGAFS analysis of deionized water samples that were spiked with $MMA^{III}$ and were stored for 5 months at 25 (a), 4 (b), and -20 °C (c), respectively. (*Source:* Gong et al. 2001.)

*Notes:* Separation was carried out on an ODS-3 column with a mobile phase containing 5 mM tetrabutylammonium hydroxide, 3 mM malonic acid, and 5% methanol. The HPLC column (15cm x 4.6mm, 3 μm particle size) was maintained at 50 °C. Peaks labeled 1-6 correspond to $As^{III}$, $MMA^{III}$, $DMA^{V}$, $MMA^{V}$, $DMA^{III}$, and $As^{V}$, respectively.

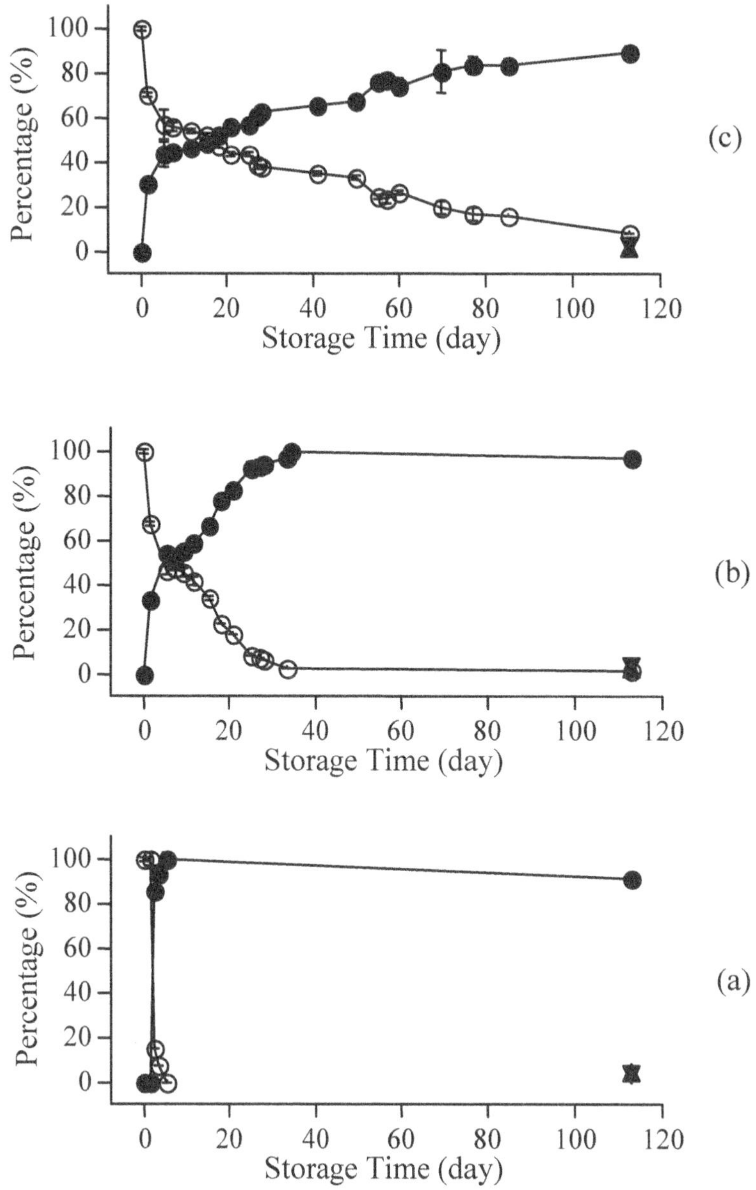


Figure 6.8. Effect of storage duration and temperature on the stability of $MMA^{III}$ in urine. (*Source:* Gong et al. 2001)

*Notes:* (a), 25 °C; (b), 4 °C; and (c), -20 °C.
(**O**), $MMA^{III}$; (●), $MMA^{V}$.

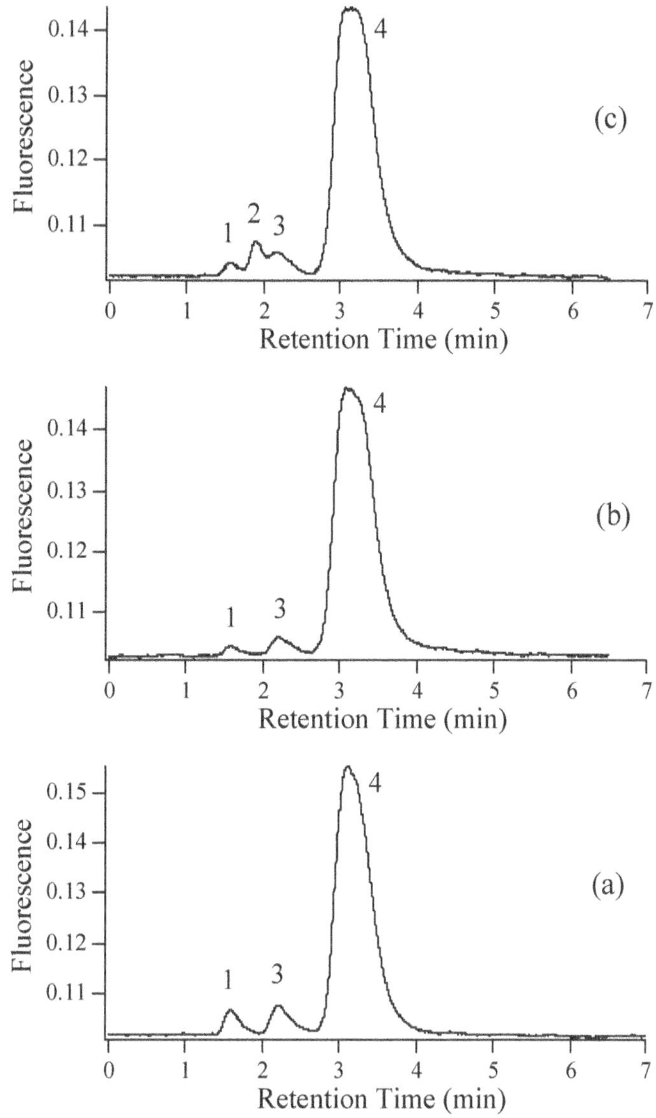


Figure 6.9. Chromatograms from the HPLC-HGAFS analysis of urine samples that were spiked with $MMA^{III}$ and were stored for 5 months at 25 (a), 4 (b), and –20 °C (c), respectively. (*Source:* Gong et al. 2001)

*Notes:* Separation was carried out on an ODS-3 column with a mobile phase containing 5 mM tetrabutylammonium hydroxide, 3 mM malonic acid, and 5% methanol. The HPLC column (15cm x 4.6mm, 3 μm particle size) was maintained at 50 °C. Peaks labeled 1-6 correspond to $As^{III}$, $MMA^{III}$, $DMA^{V}$, $MMA^{V}$, $DMA^{III}$, and $As^{V}$, respectively.

### Stability of $DMA^{III}$ in deionized water

$DMA^{III}$ in solutions was found to be very unstable under all the temperature conditions tested (Figure 6.10). Although there were variations among the storage conditions, none of these storage temperatures provided the stability needed for arsenic speciation studies. When kept at 25 °C, $DMA^{III}$ was completely oxidized to $DMA^{V}$ after 10 days. When stored at either -20 or 4 °C, complete conversion of $DMA^{III}$ to $DMA^{V}$ took 15 and 13 days, respectively. Reanalyses of these deionized water samples stored at 25 °, 4 ° and -20 °C after three months confirmed that the only detectable arsenic species was $DMA^{V}$, a result of complete oxidization of $DMA^{III}$.

### Stability of $DMA^{III}$ in urine

$DMA^{III}$ in urine is extremely unstable (Figure 6.11). When stored at 25 °C, $DMA^{III}$ in urine was completely oxidized to $DMA^{V}$ after 90 min. Additional urine samples were collected on different days and were spiked with $DMA^{III}$ and when tested, confirmed the instability of $DMA^{III}$. The conversion of $DMA^{III}$ to $DMA^{V}$ was complete after 17 and 12 h for urine samples stored at –20 °C and 4 °C, respectively. These results are consistent with the fact that $DMA^{III}$ has not been commonly detected in human urine samples (Foà et al. 1984; Buchet and Lauwerys 1994; Hakala and Pyy 1995; Hopenhayn-Rich et al. 1996b; Kalman et al. 1990; Ng, et al. 1998; Kavanagh et al. 1998; Vahter 1999; Chappell, Abernathy, and Calderon 1999), presumably because most of the $DMA^{III}$ was oxidized to $DMA^{V}$.

Qualitatively, the addition of electron donating $CH_3$- groups to $As^{III}$ results in increased electron density on $As^{III}$, and an increased tendency to be oxidized, accounting for the observed decreasing stability from $As^{III}$ to $MMA^{III}$ and $DMA^{III}$. The observation that $DMA^{III}$ is less stable to oxidation than $MMA^{III}$ is consistent with previous reports where $MMA^{III}$ was found more frequently than $DMA^{III}$ in urine samples (Aposhian et al. 2000a, 2000b; Le et al. 2000a, 2000b; Del Razo et al. 2001).

$DMA^{III}$ may have been oxidized to $DMA^{V}$ in the bladder because the present study shows that $DMA^{III}$ in urine can be readily oxidized to $DMA^{V}$ within hours. It would be of toxicological interest to study what effects $MMA^{III}$ and $DMA^{III}$ may have on the bladder since arsenic exposure is a risk factor for bladder cancer.

### Preservation of $MMA^{III}$ and $DMA^{III}$

We have made much effort to stabilize $MMA^{III}$ and $DMA^{III}$ using various potential preservatives. We found that the addition of a complexing agent, diethylammonium diethyldithiocarbamate (DDDC), was able to improve the stability of $MMA^{III}$ and $DMA^{III}$. Urine samples spiked with $MMA^{III}$ and $DMA^{III}$ were stored at -20 °C for up to 4 months and were periodically analyzed. In the presence of DDDC, $MMA^{III}$ was found to be stable for 4 months. $DMA^{III}$ was partially preserved. Approximately 80% of $DMA^{III}$ remained after storage of 3 weeks and 20% remained after 4 months. In the absence of the preservative, $DMA^{III}$ was completely oxidized to $DMA^{V}$ after 4 months. More than 50% of the $MMA^{III}$ was oxidized in the absence of the preservative. Therefore, the use of DDDC as a preservative is promising for stabilizing $MMA^{III}$ and $DMA^{III}$ for a period that is suitable for typical toxicological and epidemiological studies.

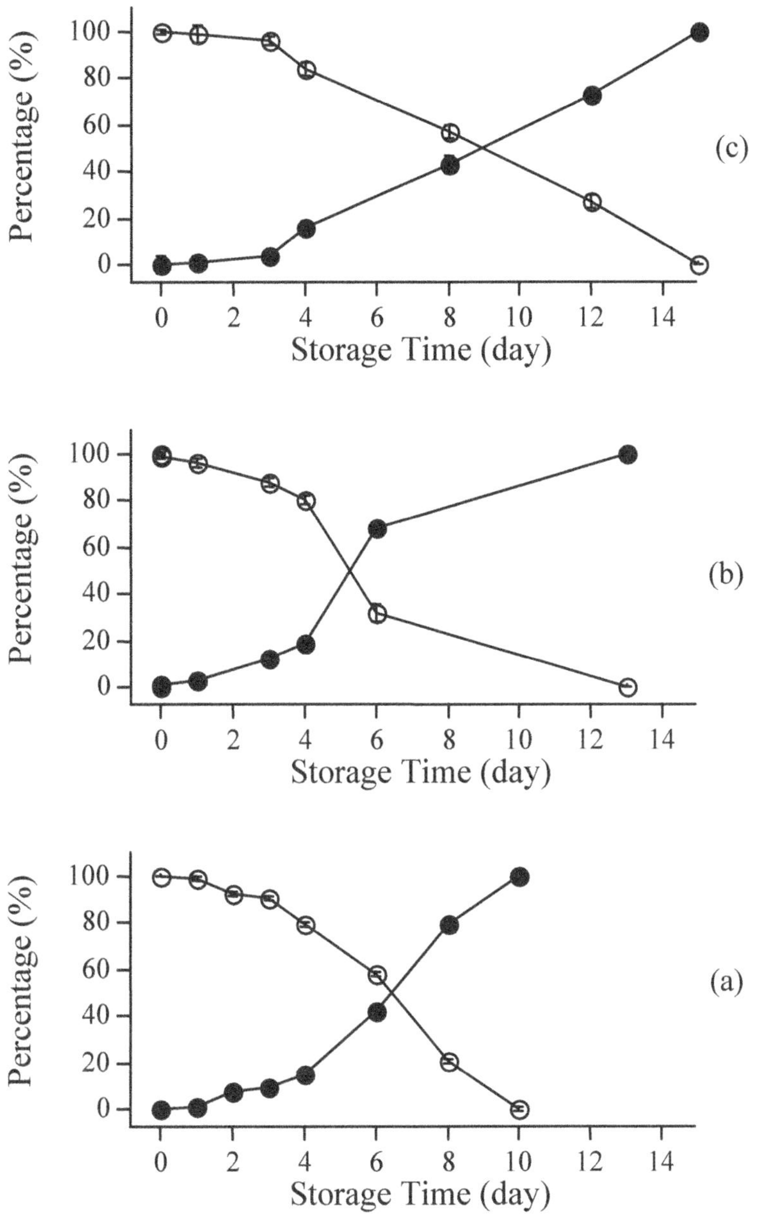


Figure 6.10. Effect of storage duration and temperature on the stability of $DMA^{III}$ in deionized water. (*Source:* Gong et al. 2001)

*Notes*: (a), 25 °C; (b), 4 °C; and (c), -20 °C.
(**O**), $DMA^{III}$ and (●), $DMA^{V}$

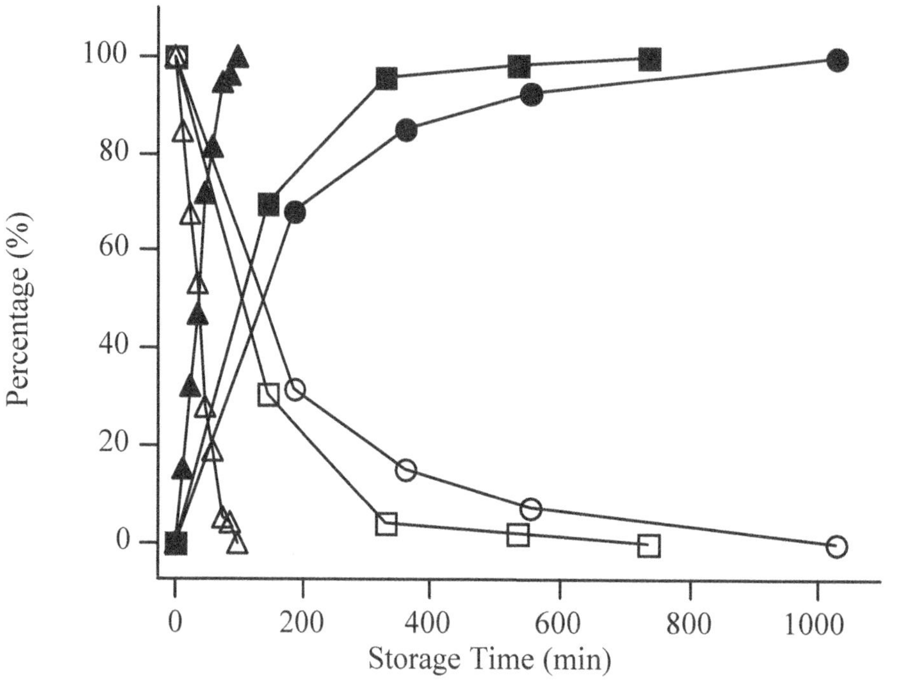


Figure 6.11. Effect of storage duration and temperature on the stability of $DMA^{III}$ in urine. (*Source:* Gong et al. 2001)

*Notes:* 25 °C: (Δ): DMA(III); (▲): DMA(V)
4 °C: (□): DMA(III); (■): DMA(V)
-20 °C: (O ): DMA(III); (●): DMA(V)

# CHAPTER 7
# INTERACTION OF ARSENICALS WITH METALLOTHIONEIN

## INTRODUCTION

The recent discovery of trivalent arsenic metabolites, monomethylarsonous acid ($MMA^{III}$) and dimethylarsinous acid ($DMA^{III}$), contributes to a better understanding of arsenic biomethylation (Le et al. 2000a). Toxicological evaluations of these reactive metabolites suggest that $MMA^{III}$ and $DMA^{III}$ are as toxic as, or more toxic than, inorganic arsenite ($As^{III}$) (Styblo et al. 1997; Lin, Cullen, and Thomas 1999; Petrick et al. 2000; Petrick et al. 2001; Mass et al. 2001; Cullen et al. 1989). However, mechanisms of action responsible for their toxicity are not clear. Trivalent arsenic binding to proteins (enzymes) has been presumed to be a cause of toxicity. However, no arsenic-containing protein has been identified.

Arsenic has been shown to induce metallothionein (MT), a stress induced protein. Several heavy metals and stress factors can induce this protein (Klaassen et al 1999; Klaassen and Liu 1998; Kreppel et al. 1993). MT is induced as a response to toxicity. Trivalent arsenic is presumed to bind with sulfulhydryl groups. Few studies have shown the interactions between arsenicals and proteins in *in vitro* systems (Bogdan et al. 1994, Styblo and Thomas 1997). However, no specific or unique arsenic-binding protein has been characterized. The objective of this study is to examine arsenic binding to proteins, using MT as an example.

## MATERIALS AND METHODS

### Reagents

Metallothionein II ($Cd_4Zn_3MT$) was purchased from Sigma (St. Louis, Mo., USA) and was used without further purification. The protein stock solution (5 mg/mL) was prepared by dissolving 5 mg metallothionein in 1 mL water. Working solutions were prepared by serial dilutions of the stock solution with water (or methanol as required). Arsenite stock solution (16.5 mM) was prepared by dissolving arsenic trioxide (Ultrapure, 99.999%, Aldrich) in water.

The source of $MMA^{III}$ was the solid oxide ($CH_3AsO$), and that of $DMA^{III}$ the iodide [$(CH_3)_2AsI$]. The precursors were prepared following literature procedures (Cullen et al. 1989, Burrows and Turner 1920), and were kept at 4 or -20 °C. Dilute solutions of the precursors were prepared fresh in deionized water to form $CH_3As(OH)_2$ ($MMA^{III}$) and $(CH_3)_2AsOH$ ($DMA^{III}$), respectively. Analyses of freshly prepared $MMA^{III}$ and $DMA^{III}$ solutions (100 μg/L) showed that their purities were approximately 98% and 97%, respectively. There was a small amount (~2%) of $As^{III}$ and $MMA^{V}$ in the $MMA^{III}$ solution. Approximately 3% $DMA^{V}$ was present in the $DMA^{III}$ reagent. TMAO was obtained from Tri Chemical Laboratory (Japan). Solutions of other standard arsenic compounds, $As^{III}$, $As^{V}$, $DMA^{V}$ (Aldrich, Milwaukee, Wis.) and $MMA^{V}$ (Chem Service, West Chester, Pa.) were prepared by appropriate dilutions with deionized water from 1000 mg/L stock solutions, as described previously (Feldmann et al., 1999, Le and Ma, 1998, Le et al. 2000a, Le et al. 1996). Formic acid, water and methanol (all HPLC grade) were purchased from Fisher Scientific (Fair Lawn, N.J.).

## Instrumentation

*HPLC-ICPMS*

A Perkin-Elmer 200 series HPLC system (PE Instruments, Shelton, Conn.) was used. BioSep-SEC 2000 size exclusion column (300x4.6 mm, Phenomenex, Torrance, Calif.) was used for separation of MT-bound from the unbound arsenicals. The chromatographic separation was carried out using isocratic elution with a mobile phase of 5 mM sodium phosphate (pH 7.0) and a flow rate of 0.8 mL/min. The injection volume was 20 μL. The effluent coming from HPLC was directly detected using an Elan 6100 DRC $^{plus}$ ICPMS (PE Sciex, Concord, Ont.), with a Turbochrom Workstation v.6.1.2 software (PE Instruments) for data processing. The operating parameters of ICP-MS were optimized to be RF power (1150 W), plasma gas flow (15 L/min), auxiliary gas flow (1.2 L/min), and nebulizer gas flow (0.9 L/min). Arsenic and cadmium were detected using the peak-hopping mode of the ICP-MS at m/z of 75 and 114, respectively.

*Triple Quadrupole Time-of-Flight Mass Spectrometry*

Mass spectrometry experiments were performed on an Applied Biosystem/MDS Sciex QSTAR Pulsar i mass spectrometer (Concord, Ont.), equipped with a Turbo Ionspray ionization source. Analyte solutions were introduced into the source by a 1-mL gas tight syringe from Hamilton (Reno, Nev.) and an integrated syringe pump, with a methanol/water mixture (50/50, v/v) and a flow rate of 10 μL/min. The mass spectrometer was operated in the positive ion mode. The instrument was calibrated daily with a commercial calibration standard. Analyst QS software (Applied Biosystems, Foster City, Calif.) was used for the spectrum acquisition and data analysis. IgorPro software (WaveMetrics, Lake Oswego, Ore.) was used to plot the spectra.

In single MS scan mode, the mass measurements (900-2000 amu) were performed using the time-of-flight (TOF) section of the instrument, while Q1 and Q2 were operated in the RF-only mode. Mass spectra were acquired with an electrospray voltage of 5500 V, first declustering potential (DP1) of 65 V, second declustering potential (DP2) of 15 V and focusing potential (FP) of 215 V. The resolution was 10,000 (fwhm) at m/z 850 with a 10-ppm mass accuracy using internal standard.

In the MS/MS mode, the parent ion was selected by the first quadrupole (Q1) with low mass resolution, and fragmented in the second quadrupole (Q2) by collision-induced dissociation (CID) with a collisional energy of 90 eV and collision gas setting of 12. The resulting product ions were analyzed by TOF analyzer. A four-anode detector was used with a time-to-digital converter capable of detecting a single ion. In assigning these product ions, the theoretical fragmentation pattern of MT was generated for comparison using the MS-product tool in Protein Prospector (http://prospector.ucsf.edu).

## Procedures

$As^{III}$, $MMA^{III}$, and $DMA^{III}$ (1 μM) in deionized water were each incubated with various concentrations of metallothionein (5-15 μM) in phosphate buffered saline (PBS) for 1-2 h. These solutions were subject to HPLC-ICPMS analysis of MT-As complexes. The MT-bound and free arsenic species were separated on a size exclusion column with phosphate buffer (pH 7.3) as a mobile phase, and were detected using the ICPMS. Both arsenic (m/z 75) and cadmium (m/z

114) were detected simultaneously to confirm the binding of arsenic with MT because cadmium was present in the native MT.

To study the binding stoichiometry between MT and $As^{III}$, $MMA^{III}$, and $DMA^{III}$ species, a constant concentration of MT (7 μM) was incubated with varying concentrations of the trivalent arsenic species to obtain arsenic to MT molar ratios of 1:20, 1:5, 1:1, 5:1, 20:1, and 100:1. The mixture solutions were incubated at room temperature for 2 h. They were diluted with 50% methanol immediately prior to electrospray MS analysis. The concentration of MT in the dilute solutions was ~7 μM, and the concentrations of the arsenic species ranged from 0.35 to 700 μM.

Initially, the MT/arsenic mixture solutions were acidified with 1% formic acid to remove Zn and Cd from the native MT and to simplify the mass spectra of MT-As complexes. Subsequent experiments were carried out without the acidification. The incubation of MT with arsenic species under neutral pH conditions, followed by electrospray MS and HPLC-ICPMS analysis, confirmed the presence of the complexes between MT and the trivalent arsenic species.

## RESULTS AND DISCUSSION

### Binding of Arsenic to Metallothionein

To demonstrate that trivalent arsenicals bind to MT, we first developed a technique that is based on size exclusion HPLC separation of the MT-bound arsenic from the unbound arsenic followed by specific detection of arsenic using ICPMS. Figure 7.1 shows typical chromatograms from the HPLC/ICPMS analysis of solutions containing MT and $As^{III}$ and $MMA^{III}$. The MT-bound $As^{III}$ and MT-bound $MMA^{III}$ are clearly separated from the unbound $As^{III}$ and $MMA^{III}$. Simultaneous detection of arsenic and cadmium provided a reference of the MT-bound fraction because cadmium is present in the native MT. These results show that trivalent arsenic can be readily bound to MT. Additional experiments carried out using $As^{III}$, $MMA^{III}$, and $DMA^{III}$ (1 μM) and MT of various concentrations (1-50 μM) confirmed the binding.

To further understand the binding of arsenicals to MT and their binding stoichiometry, we used electrospray ionization mass spectrometry (ESI/MS) to examine various species resulting from the binding. The native rabbit MT contained both Zn and Cd. Analyses of MT solutions by ESI/MS revealed that the binding of Zn and Cd with MT was pH dependent. At pH 6, both Zn and Cd remained intact on the MT. When the pH of the MT solution was decreased to 4.5, zinc was released, leaving only Cd on the MT. When the pH of the solution was further lowered to 2.0, Cd was also released from the protein, resulting in apo-MT. Characteristic ions (m/z) for the apo-MT detected at pH 2.0 included 1022.0 ($6^+$-charge state), 1226.2 ($5^+$-charge state), and 1532.5 ($4^+$-charge state). These correspond to a molecular mass of 6126 for the apo-MT, consistent with the expected values (6126.3). In addition, sodium adducts of MT were observed at m/z of 1025.6 ($6^+$), 1230.6 ($5^+$) and 1538.0 ($4^+$) as expected.

To simplify the interpretation of mass spectra, initial studies of MT binding with arsenicals were carried out using apo-MT that was obtained by acidifying the native MT solution to pH 2.0 with 0.5% formic acid. Figure 7.2 shows typical mass spectra obtained from solutions containing 7 μM MT and increasing concentrations of $As^{III}$ (0.35-140 μM). The molar ratios of $As^{III}$ to MT ranged from 1:5 to 20:1. Multiply-charged ions ($5^+$ and $4^+$) were predominant. With the increase of $As^{III}$ concentration at a constant concentration of MT, the number of $As^{III}$ atoms bound to the MT increased.

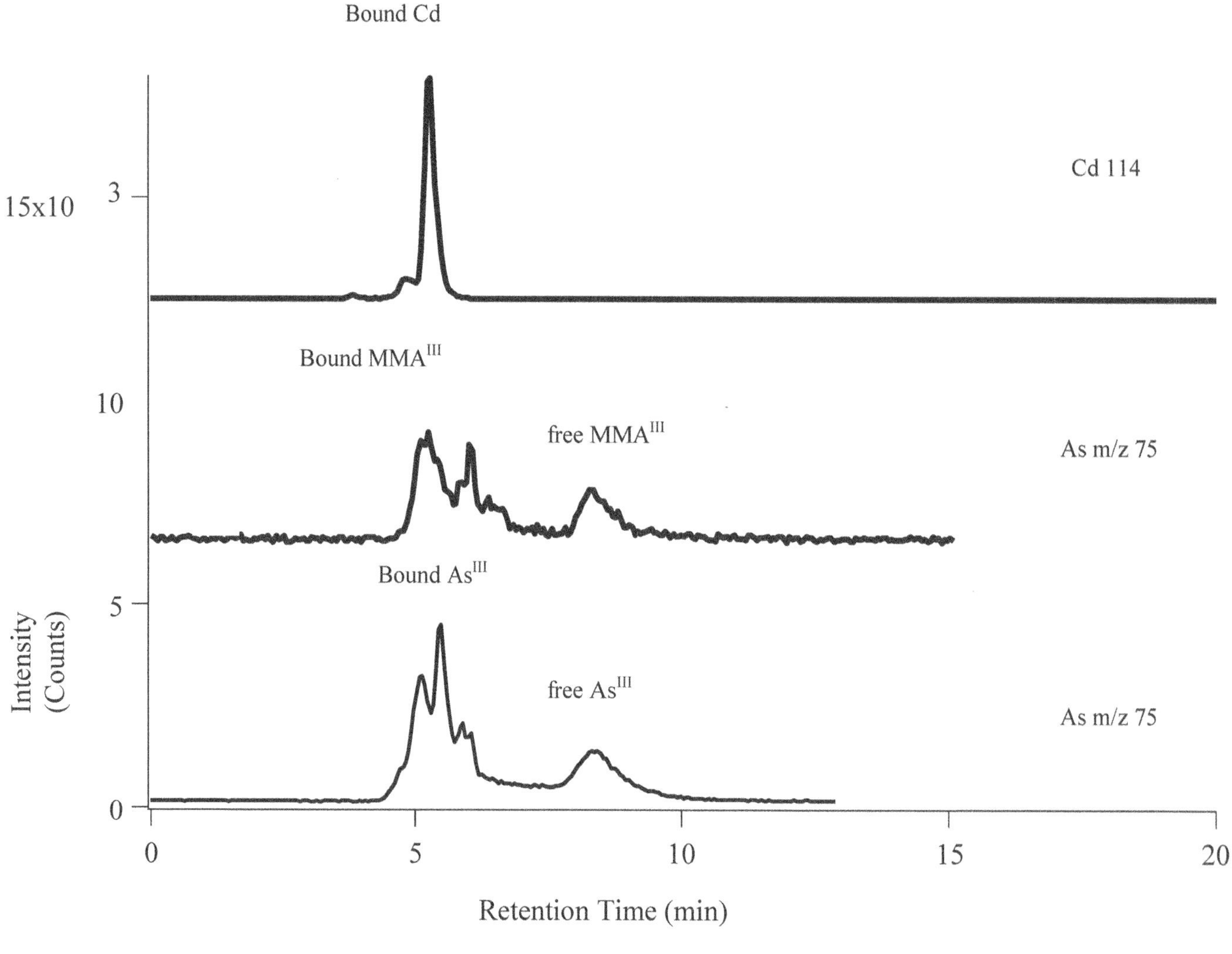


Figure 7.1. HPLC/ICPMS analysis of reaction mixtures containing MT and $As^{III}$ or $MMA^{III}$ showing the presence of MT-$As^{III}$ and MT-$MMA^{III}$ complexes. (*Source:* Jiang et al. 2003a)

The numbers marked on the peaks represent the number of arsenic atoms bound to the MT. At a low concentration of $As^{III}$ relative to MT (Figure 7.2a), the apo-MT ions, m/z 1226.2 carrying $5^+$ charges and 1532.5 carrying $4^+$ charges, were observed as the dominant species. Complex peaks corresponding to the apo-MT bound with one arsenic (peak 1) (m/z 1240.7 carrying $5^+$ and 1550.8 carrying $4^+$) were also observed. When the $As^{III}$ concentration was increased to equal the concentration of MT, the MT-As (peak 1), MT-$As_2$ (peak 2), MT-$As_3$ (peak 3), and MT-$As_4$ (peak 4) complexes along with their sodium adducts were observed (Figure 7.2b). MT-$As_2$ complex (peak 2) was the most abundant species. In the presence of 5-fold excess of $As^{III}$ over MT, MT complexes with 5 and 6 arsenic moieties were formed (Figure 7.2c). Increasing $As^{III}$ to an As:MT ratio of 20:1 resulted in the formation of the MT-$As_6$ complex as the dominant species (Figure 7.2d). It appears that the maximum number of arsenic atoms bound to the MT was 6. A further increase of $As^{III}$ to an As:MT ratio of 100:1 did not show the formation of MT complex with more than 6 $As^{III}$ moieties.

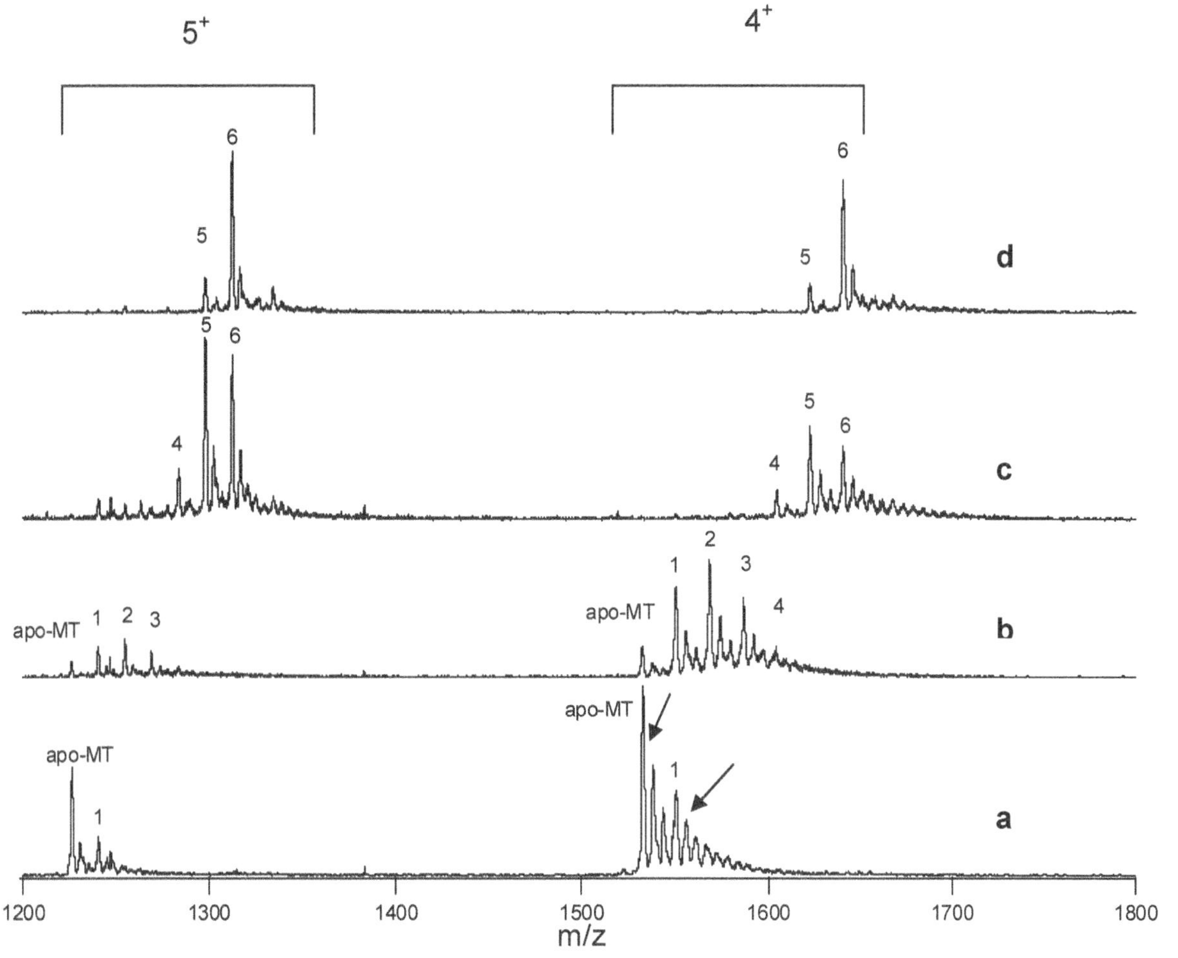


Figure 7.2. ESI mass spectra from the analysis of solutions containing 7 μM MT and varying amounts of $As^{III}$ (0.35, 7, 35, and 140 μM). MT (7 μM) and $As^{III}$ (0.35-140 μM) in deionized water were incubated at room temperature for 2 h. The solution was diluted with 50% methanol and acidified with formic acid to pH 2.0 immediately prior to ESI/MS analysis. The peaks labeled with numbers were compexes of MT and $As^{III}$. The numbers on the peaks represent the number of $As^{III}$ bound to the MT molecule. For example, peak 6 represent MT-$As_6$. The peaks labeled with arrows are sodium adducts of the apo-MT and MT-As complexes. Not all of the sodium adducts are labeled. The ratios of $As^{III}$ to MT are (a) 1:5; (b) 1:1; (c) 5:1; and (d) 20:1. (*Source:* Jiang et al. 2003a)

The binding of each MT molecule with 6 $As^{III}$ moieties in the presence of excess $As^{III}$ is understandable as schematically illustrated in Figure 7.3. MT contains 20 cysteine residues. Because each $As^{III}$ is able to bind with three cysteines, the maximum number of $As^{III}$ that can be bound on a MT molecule is 6. Results in Figure 7.2 are consistent with arsenic coordination chemistry. Further studies using $MMA^{III}$ and $DMA^{III}$ binding with MT confirmed the arsenic binding stoichiometry.

The assignments of the mass spectral peaks were supported with the accurate mass measurements of the species. Table 7.1 summarizes the theoretical and the measured masses of the MT and the $MT(As)_6$ species. The measured molecular mass of each species was an average obtained from triplicate measurements of the $5^+$ and $4^+$ ions. They were obtained without using an internal standard for mass calibration. The excellent match between the measured and expected values support the assignment of the peaks.

Table 7.1
Theoretical and measured molecular masses of the MT and MT-$As^{III}$ species

| Species | Theoretical values (Da) Mol mass | Experimental values | | | | |
|---|---|---|---|---|---|---|
| | | $5^+$ charge state (m/z) | $4^+$ charge state (m/z) | Mol mass (Da) | Δm (Da) | Mass accuracy (ppm) |
| MT | 6126.3 | 1226.20 | 1532.51 | 6126.01 | -0.29 | -46 |
| MT-As | 6198.20 | 1240.66 | 1550.76 | 6198.67 | 0.47 | 76 |
| MT-$As_2$ | 6270.10 | 1254.74 | 1568.38 | 6269.10 | 1.0 | -159 |
| MT-$As_3$ | 6341.99 | 1269.15 | 1586.74 | 6341.85 | -0.14 | -22 |
| MT-$As_4$ | 6413.89 | 1283.97 | 1604.73 | 6414.88 | 0.99 | 154 |
| MT-$As_5$ | 6485.79 | 1298.18 | 1622.48 | 6485.91 | 0.12 | 18 |
| MT-$As_6$ | 6557.69 | 1312.65 | 1640.74 | 6558.61 | 0.92 | 139 |

*Source:* Jiang et al. 2003a

The presence of MT-As complexes was further confirmed with analyses of fragment ions of MT-As species by using tandem mass spectrometry. Collision-induced dissociation of the MT-As species resulted in arsenic-containing fragments. Table 7.2 shows arsenic-containing fragment ions detected from the dissociation of $MT(As)_6$ species ($5^+$ ion at m/z 1640.7, peak 6 in Figure 7.2d) with a collision cell energy of 90 eV. The presence of arsenic-cysteine fragments suggests that $As^{III}$ is bound to the cysteine residues in MT. In addition to these arsenic-containing fragments, other fragment ions were consistent with the theoretical fragmentation pattern of MT protein that was generated by using MS-product tool in prospector (http://prospector.ucsf.edu). With a lower collision cell energy below 40 eV, little fragmentation was observed and the parent ion (MT-$As_6$) was the most abundant species, suggesting the strong binding of $As^{III}$ with MT.

Table 7.2
Expected and experimental masses of arsenic related amino acids species observed in tandem mass spectra.

| | Expected values (Da) | Experimental values (Da) | Δm (Da) | Mass accuracy Δm/m (ppm) |
|---|---|---|---|---|
| As-$SCH_2CHN^+$ | 147.920 | 147.922 | 0.002 | 14 |
| As-$SCH_2CHNH_2^+$ | 149.936 | 149.935 | -0.001 | 7 |
| As-**CysGly** | 234.954 | 234.960 | 0.006 | 26 |
| As-**CysArg** | 248.964 | 248.968 | 0.004 | 16 |
| As-**CysSer** | 264.964 | 264.977 | 0.013 | 47 |
| As-**CysTyr** | 278.974 | 278.964 | -0.010 | 36 |
| As-**CysGlyArg** | 305.994 | 305.992 | -0.002 | 7 |
| As-**CysSerArg** | 336.004 | 335.981 | -0.023 | 68 |

*Source:* Jiang et al. 2003a
Note: The letters in bold stand for the amino acids.

We carried out further experiments to confirm that the formation of the MT-As complexes in solution was real, and not an artifact of the electrospray process. We examined the formation of MT-As complexes over different incubation periods in solution and the results indicated that the complex formation was dependent on the incubation in solution. When MT and $As^{III}$ (4:1 molar ratio) was mixed and immediately analyzed using ESI/MS, only MT (1226.2 at $5^+$ and 1532.5 at $4^+$ charges) and MT-$As_1$ (1240.7 at $5^+$ and 1550.8 at $4^+$ charges) species were observed. Repeat ESI/MS analysis of the same mixture after an hour incubation demonstrated the presence of MT-$As_5$ and MT-$As_6$ species (data not shown). These results suggest that the formation of MT-As complexes took place in the liquid phase and depended on the reaction time in solution.

## Binding of $MMA^{III}$ and $DMA^{III}$ to Metallothionein

Having established the binding of MT with $As^{III}$, we further examined the binding of MT with the trivalent arsenic metabolites, $MMA^{III}$ and $DMA^{III}$, to gain further details on the binding stoichiometry. Inorganic $As^{III}$ is able to bind to three thiol groups, forming $As(\text{-Cys})_3$. Therefore, up to 6 $As^{III}$ could be bound to a single MT as there are 20 cysteine residues available in the MT. This is evident from Figures 7.2 and 7.3. $MMA^{III}$ [$CH_3As(OH)_2$] would be able to bind with two thiols. $DMA^{III}$ [$(CH_3)_2AsOH$] would be able bind to only one thiol group. Therefore, up to 10 $MMA^{III}$ and 20 $DMA^{III}$ could be bound to each MT when $MMA^{III}$ and $DMA^{III}$ are in excess. Figures 7.4 and 7.5 clearly demonstrate this binding stoichiometry.

At a $MMA^{III}$ to MT ratio of 1:5, the apo-MT is the dominant species (Figure 7.4 trace a). No MT-$MMA^{III}$ complex was evident. Increasing the $MMA^{III}$ concentration to equal the molar concentration of MT resulted in the formation of several MT-$MMA^{III}$ complex species. MT-[$As(CH_3)$], MT-[$As(CH_3)$]$_2$, MT-[$As(CH_3)$]$_3$, MT-[$As(CH_3)$]$_4$ and MT-[$As(CH_3)$]$_5$ are clearly observed (Figure 7.4, trace b). Further increasing the $MMA^{III}$ concentration to a 5-fold excess over MT leads to the binding of a maximum of ten $MMA^{III}$ moieties on the MT, MT-[$As(CH_3)$]$_{10}$ (Figure 7.4, trace c). With a 50-fold excess of $MMA^{III}$ over MT, the only complex species detected is MT-[$As(CH_3)$]$_{10}$ (Figure 7.4, trace d).

The experimental masses of MT-$[As(CH_3)]_n$ are listed in Table 7.3. The mass difference between the MT-$[As(CH_3)]_n$ complex and the apo-MT is 87.9n, where n is the number of $MMA^{III}$ moieties attached to MT and varies from 1 to 10. This mass difference is consistent with the addition of the expected mass of 87.9294 for $AsCH_3$ and the loss of 2H from the thiol groups. This is due to the formation of $CH_3As$ complex with MT through binding with two thiols (resulting in loss of the proton from each thiol group) as illustrated in Figure 7.3.

Figure 7.5 shows spectra from the analysis of $DMA^{III}$ binding to MT. Four spectra were acquired from reaction mixtures containing four different ratios of $DMA^{III}$ to MT. The complex peaks corresponding to MT-$[As(CH_3)_2]_n$ are observed in these spectra, in which n varies from 1 to 20. The observed maximum number of $As(CH_3)_2$ binding MT is 20 even with a MT to DMA ratio of 1: 200 in solution (data not shown). The observed masses and expected masses for MT-$[As(CH_3)_2]_n$ are listed in Table 7.4. The measured mass differences between the MT-$DMA^{III}$ complexes and the apo-MT are 103n, where n is the number of $DMA^{III}$ moieties attached to MT and varies from 1 to 20. This is consistent with the addition of the expected mass of 103.9607 for $As(CH_3)_2$ and the loss of a proton from a thiol group.

$MMA^{III}$ and $DMA^{III}$ are two crucial metabolites of arsenic biomethylation process. The use of $As^{III}$, $MMA^{III}$, and $DMA^{III}$ is a unique system, allowing us to clearly illustrate the binding stoichiometry.

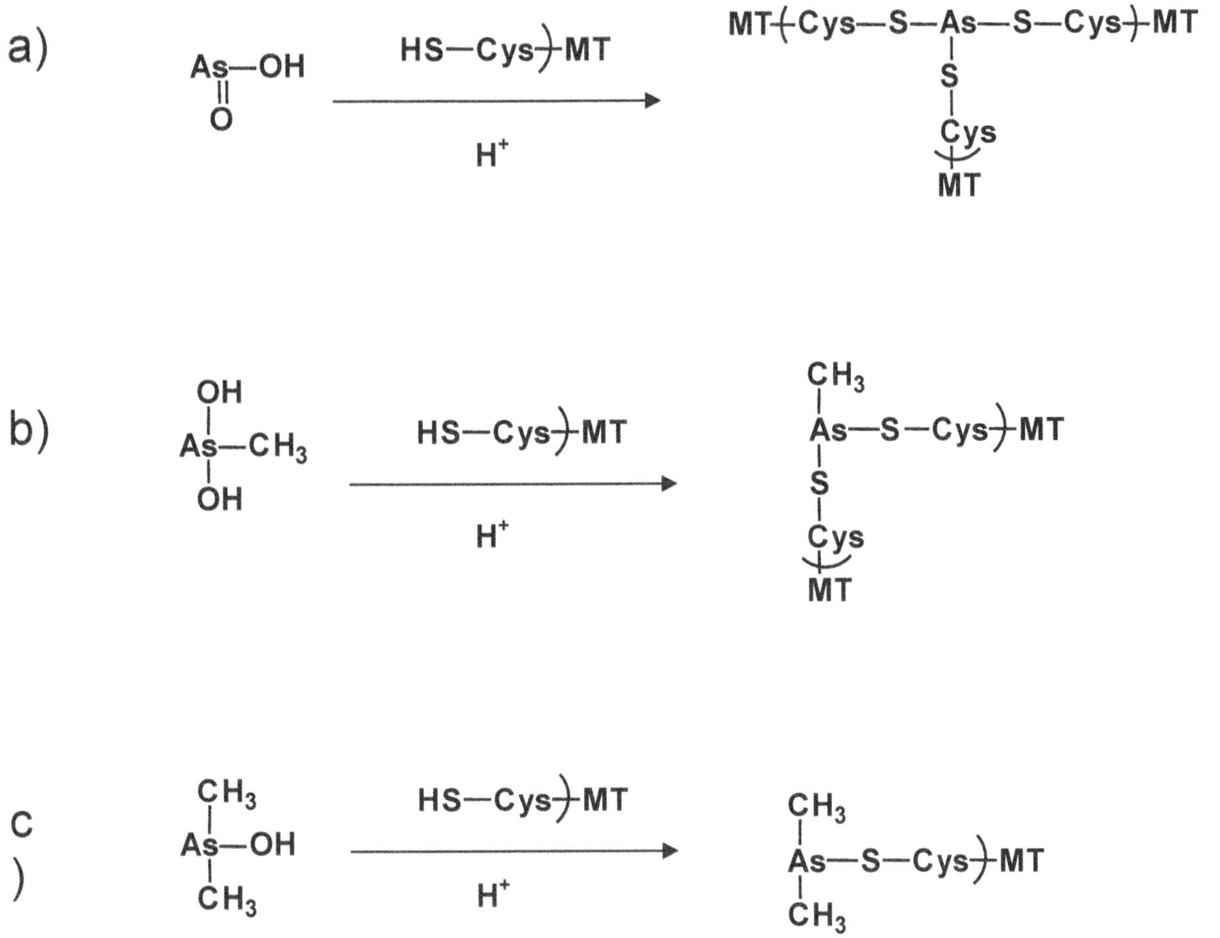

Figure 7.3. Schematic representation of the binding stoichiometry between MT and $As^{III}$ (a), $MMA^{III}$ (b), and $DMA^{III}$ (c). (*Source:* Jiang et al. 2003a)

Figure 7.6 shows tandem mass spectra of MT-$MMA^{III}$ and MT-$DMA^{III}$ by selecting $MT[As(CH_3)]_7$ at m/z of 1686.4 ($4^+$) and $MT[As(CH_3)_2]_7$ at m/z of 1714.0 ($4^+$) as the parent ion, respectively. The peak observed at m/z of 163.9573 for MT-$MMA^{III}$ in Figure 7.6 corresponds to $CH_3AsSC_2H_4N^+$ (163.9515), resulting from $CH_3As$ and cysteine binding. The $As(CH_3)_2S^+$ (136.9406) located at m/z of 136.9428 and $As(CH_3)_2SC_2H_5N^+$ (179.9828) at m/z of 179.9885 also arise from the $As(CH_3)_2$-cysteine binding. Observation of these arsenical-containing fragments further supports the binding of trivalent arsenic with MT as shown in Figure 7.3.

Table 7.3
Expected and experimental molecular masses for unbound MT and MT bound with $MMA^{III}$

| MT | Theoretical values (Da) Mol mass | Experimental values $5^+$ state (m/z) | $4^+$ state (m/z) | Mol mass (Da) | Δm (Da) | Mass accuracy (ppm) |
|---|---|---|---|---|---|---|
| apo | 6126.30 | 1226.20 | 1532.65 | 6126.30 | 0 | 0 |
| apo+[As $(CH_3)$] | 6214.23 | 1243.97 | 1554.49 | 6214.41 | 0.18 | 29 |
| apo+2[As $(CH_3)$] | 6302.16 | 1261.36 | 1576.50 | 6301.90 | -0.26 | -41 |
| apo+3[As $(CH_3)$] | 6390.09 | 1279.30 | 1598.63 | 6390.99 | 0.9 | 142 |
| apo+4[As $(CH_3)$] | 6478.02 | 1296.71 | 1620.60 | 6478.48 | 0.46 | 72 |
| apo+5[As $(CH_3)$] | 6565.95 | 1314.09 | 1642.62 | 6565.97 | 0.02 | 4 |
| apo+6[As $(CH_3)$] | 6653.88 | 1331.93 | 1664.52 | 6654.36 | 0.48 | 73 |
| apo+7[As $(CH_3)$] | 674181 | 1349.52 | 1686.46 | 6742.22 | 0.41 | 62 |
| apo+8[As $(CH_3)$] | 6829.74 | 1367.12 | 1708.62 | 6830.53 | 0.79 | 116 |
| apo+9[As $(CH_3)$] | 6917.66 | 1384.56 | 1730.59 | 6918.09 | 0.43 | 62 |
| apo+10[As $(CH_3)$] | 7005.59 | 1402.06 | 1752.39 | 7005.41 | -0.18 | -26 |

*Source:* Jiang et al. 2003b

Table 7.4
Expected and experimental molecular masses for unbound MT and MT bound with $DMA^{III}$

| MT | Theoretical values (Da) Mol mass | Experimental values $4^+$ charge state (m/z) | Mol mass (Da) | Δm (Da) | Mass accuracy (ppm) |
|---|---|---|---|---|---|
| apo | 6126.3 | 1532.51 | 6126.03 | -0.27 | -44 |
| apo+[As $(CH_3)_2$] | 6230.26 | 1558.57 | 6230.29 | 0.03 | 4 |
| apo+2[As $(CH_3)_2$] | 6334.22 | 1584.86 | 6335.43 | 1.21 | 190 |
| apo+3[As $(CH_3)_2$] | 6438.18 | 1610.89 | 6439.57 | 1.39 | 216 |
| apo+4[As $(CH_3)_2$] | 6542.14 | 1636.75 | 6543.01 | 0.87 | 132 |
| apo+5[As $(CH_3)_2$] | 6646.10 | 1662.27 | 6645.06 | -1.04 | -155 |
| apo+6[As $(CH_3)_2$] | 6750.06 | 1688.27 | 6749.06 | -1.0 | -148 |
| apo+7[As $(CH_3)_2$] | 6854.02 | 1714.03 | 6852.11 | -1.91 | -279 |
| apo+8[As $(CH_3)_2$] | 6957.99 | 1740.20 | 6956.81 | -1.18 | -168 |
| apo+9[As $(CH_3)_2$] | 7061.91 | 1765.77 | 7059.07 | -2.84 | -407 |
| apo+10[As $(CH_3)_2$] | 7165.91 | 1792.69 | 7166.74 | 0.83 | 117 |
| apo+11[As $(CH_3)_2$] | 7269.87 | 1818.60 | 7270.40 | 0.53 | 73 |
| apo+12[As $(CH_3)_2$] | 7373.83 | 1844.70 | 7374.80 | 0.97 | 132 |
| apo+13[As $(CH_3)_2$] | 7477.79 | 1870.93 | 7479.73 | 1.96 | 260 |
| apo+14[As $(CH_3)_2$] | 7581.75 | 1896.72 | 7582.87 | 1.12 | 148 |
| apo+15[As $(CH_3)_2$] | 7685.71 | 1922.78 | 7687.13 | 1.42 | 185 |
| apo+16[As $(CH_3)_2$] | 7789.67 | 1948.80 | 7791.20 | 1.53 | 196 |
| apo+17[As $(CH_3)_2$] | 7893.63 | 1974.71 | 7894.85 | 1.23 | 154 |
| apo+18[As $(CH_3)_2$] | 7997.59 | 2000.98 | 7999.93 | 2.34 | 292 |
| apo+19[As $(CH_3)_2$] | 8101.55 | 2026.71 | 8102.83 | 1.28 | 158 |
| apo+20[As $(CH_3)_2$] | 8205.51 | 2052.85 | 8207.44 | 1.87 | 231 |

*Source:* Jiang et al. 2003b

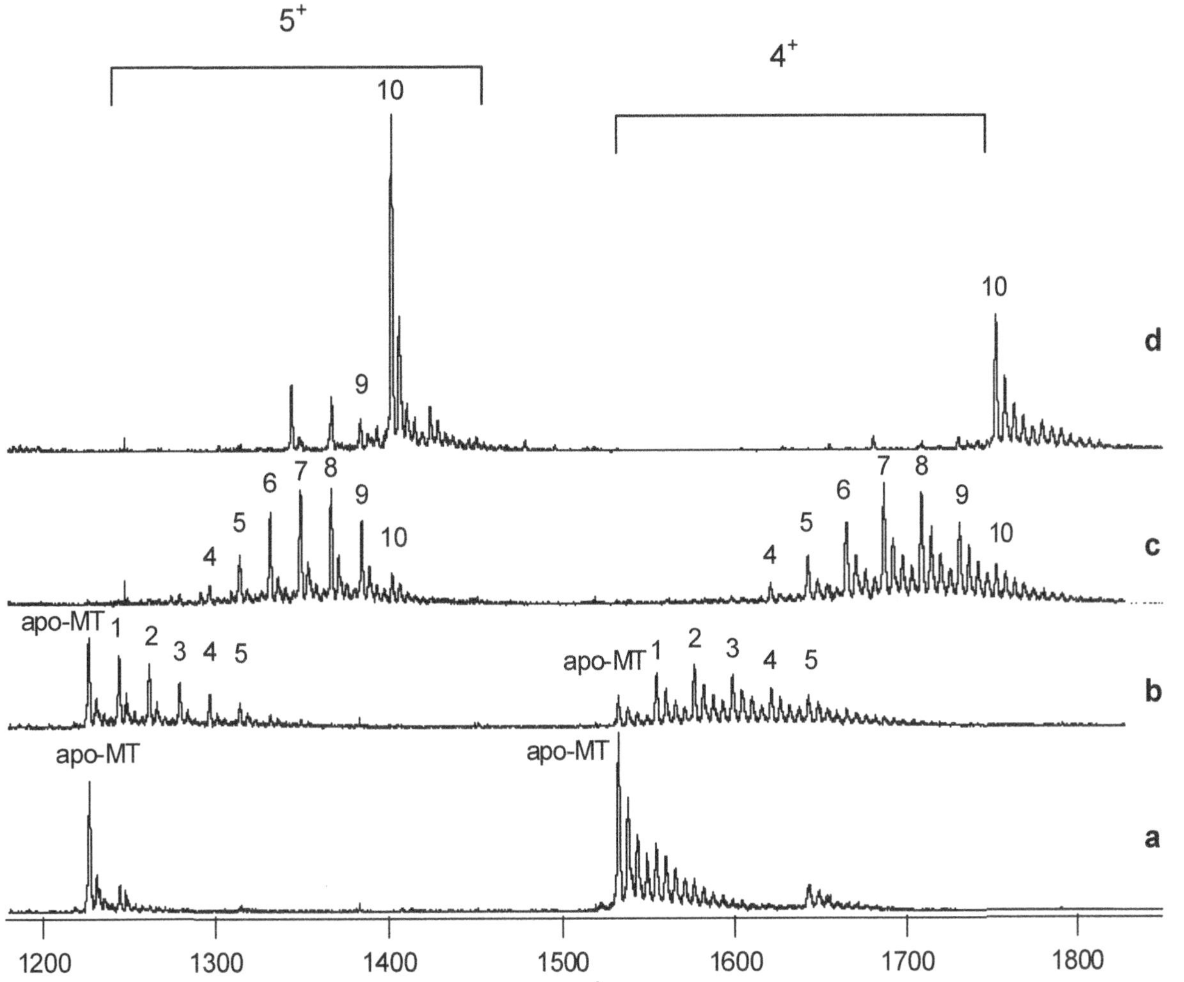


Figure 7.4. ESI mass spectra from the analysis of solutions containing 7 μM MT and varying amounts of $MMA^{III}$ (μM). MT (7 μM) and $MMA^{III}$ (0.35-350 μM) in deionized water were incubated at room temperature for 2 h. The solution was diluted with 50% methanol and acidified with formic acid to pH 2.0 immediately prior to ESI/MS analysis. The peaks labeled with numbers were complexes of MT and $MMA^{III}$. The numbers on the peaks represent the number of $As^{III}$ bound to the MT molecule. For example, peak 6 represents MT-$As_6$. The ratios of $MMA^{III}$ to MT are (a) 1:5; (b) 1:1; (c) 5:1; and (d) 50:1. (*Source:* Jiang et al. 2003a)

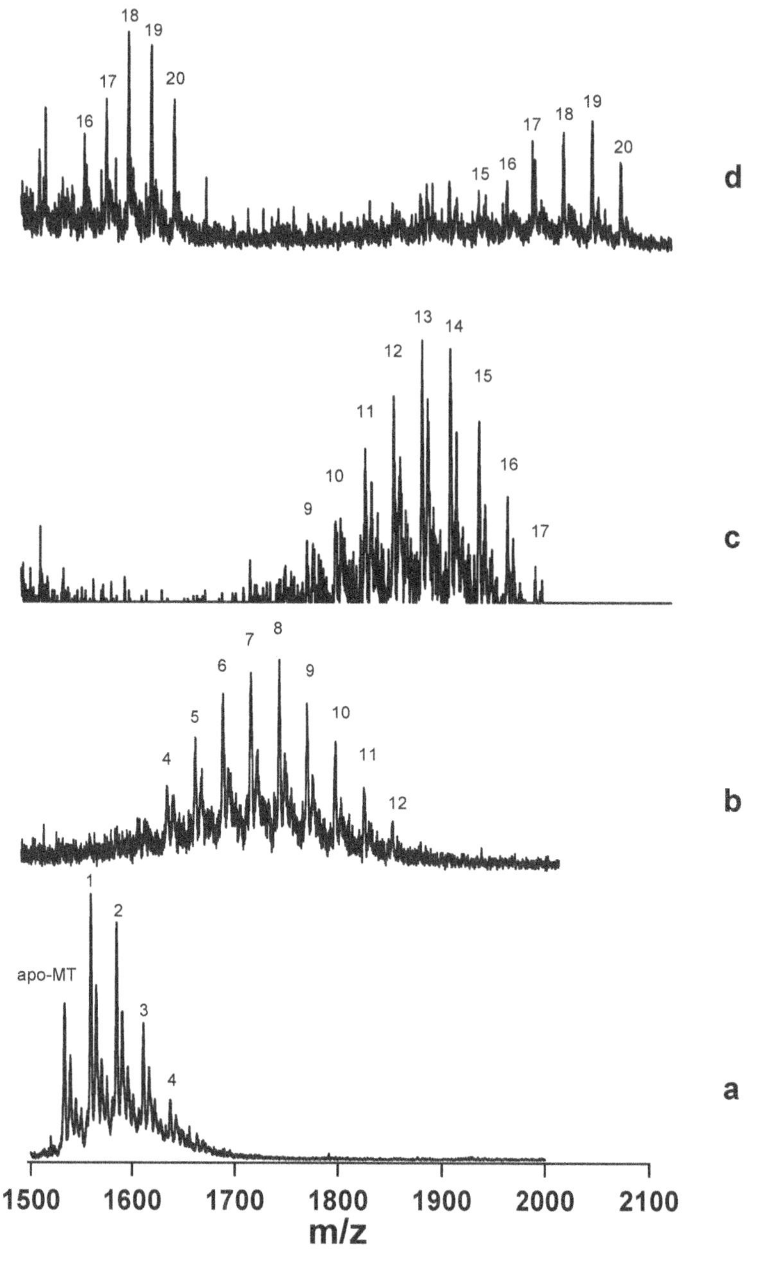


Figure 7.5. ESI mass spectra from the analysis of solutions containing 7 μM MT and varying amounts of DMA$^{III}$ (μM). MT (7 μM) and DMA$^{III}$ (0.35-350 μM) in deionized water were incubated at room temperature for 2 h. The solution was diluted with 50% methanol and acidified with formic acid to pH 2.0 immediately prior to ESI/MS analysis. The peaks labeled with numbers were complexes of MT and DMA$^{III}$. The numbers on the peaks represent the number of As$^{III}$ bound to the MT molecule. For example, peak 6 represents MT-$As_6$. The ratios of DMA$^{III}$ to MT are (a) 1:5; (b) 1:1; (c) 5:1; and (d) 50:1. (*Source:* Jiang et al. 2003a)

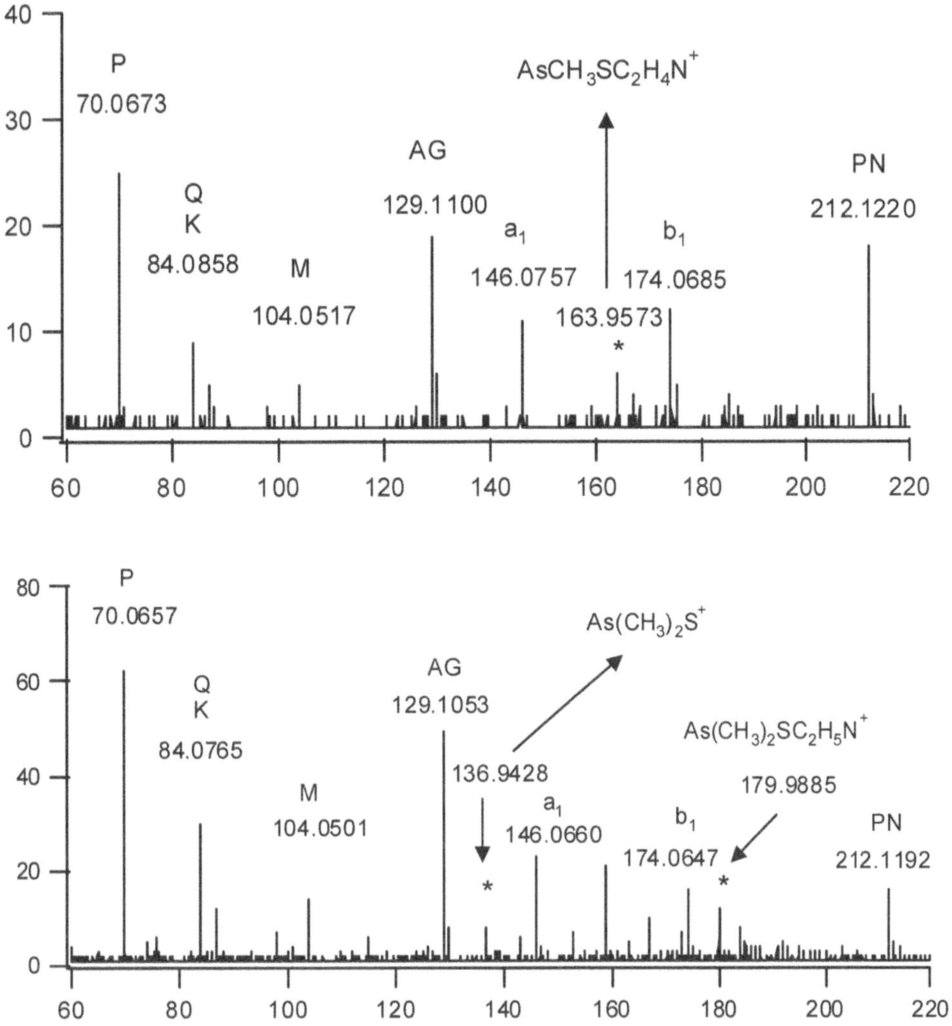


Figure 7.6. Partial ESI/MS/MS spectra showing the low mass region for MT[As($CH_3$)]$_7$ (a) and MT[As($CH_3$)$_2$]$_7$ (b). The ion at m/z of 1686.4 and 1714.0 with $4^+$-charge state were fragmented at the collision cell. The peaks at m/z of 163.9573 in a) and 136.9428, 179.9885 in b) are the arsenic related fragment ions with possible formula listed as well, according to the mass measurements. (*Source:* Jiang et al. 2003a)

Experiments were also carried out to explore whether the pentavalent arsenic species, $TMAO^V$, $DMA^V$, $MMA^V$ and $As^V$, would bind with metallothionein. $TMAO^V$, $(CH_3)_3AsO$, could be reduced by MT to $(CH_3)_3As$. However, $(CH_3)_3As$ has no available binding site for thiol groups. We only observed a shift of MT peaks to a lower mass by 20 m/z units. This corresponds to an oxidation of –SH to disulfide due to reduction of $TMAO^V$ to $TMA^{III}$ (trimethylarsine). The reduced product, trimethylarsine, has no binding site for reacting with cysteine. Consistent with arsenic coordination chemistry, our results showed no binding between $TMAO^V$ and MT.

$DMA^V$ could be reduced by the cysteines in MT, and disulfide bond between two cysteines would be formed in the oxidation. The $DMA^{III}$ that would be produced as the result of the reduction reaction would be expected to form MT-$DNA^{III}$ complexes. There was a reaction as judged by the many new peaks in the mass spectra (data not shown). However, the formation of many species from the redox reaction resulted in complicated mass spectra. We were not able to confirm the presence of any MT-DMA complex from reactions between MT and $DMA^V$. Similarly, we did not confirm the presence of MT-MMA and MT-As complexes from reactions between MT and the pentavalent $MMA^V$ and $As^V$ species.

## DISCUSSION AND CONCLUDING REMARKS

The main metabolic process that results from the uptake or ingestion of inorganic arsenic is biomethylation. The intermediate trivalent metabolites, $MMA^{III}$ and $DMA^{III}$, are at least as toxic as inorganic arsenite ($As^{III}$). The mechanism(s) of action responsible for the toxicity of the trivalent arsenic species have not been completely elucidated. Our study clearly demonstrates that $As^{III}$, $MMA^{III}$ and $DMA^{III}$ readily react with sulfhydryl groups in MT. The novel use of $As^{III}$, $MMA^{III}$ and $DMA^{III}$ for this study allowed us to clearly illustrate the binding stoichiometry. Toyama et al. (2002) have examined the interactions of $As^{III}$ with human MT-II, using UV absorption, ICP atomic emission spectrometry, and matrix-assisted laser desorption ionization mass spectrometry. Our results on $As^{III}$ binding with MT are consistent with their report: they also found that the maximum molar ratio of $As^{III}$ to MT was 6:1. They did not examine the binding of MT with other arsenic species. Our study provides further detailed information on binding stoichiometry of three trivalent arsenic species with metallothionein.

Studies of biochemical interactions between arsenic and proteins are crucial to a better understanding of arsenic health effects. $Arsenic^{III}$ has been shown to interact with glutathione to produce 3:1 glutathione:arsenic complex through thiolate sulfur atoms (Serves et al. 1995), and cysteine reacts with arsenite to produce an $As(Cys)_3$ complex (Johnson and Voegtlin 1930). Recently, Farrer et al. (2000) found that $As^{III}$ is able to distort polypeptide structure in order to satisfy its preferred trigonal-pyramidal thiolate coordination. The peptides used were synthesized by substitution of a leucine residue for cysteine at position 12 or 16 of one TRI family peptide, introducing a thiolate for metal binding. Previous studies on arsenical peptide/protein interactions (Chen, Mobley, and Rosen 1985; Ji and Silver 1992; Stevens et al. 1999) have also recognized the roles of protein in the metabolism and possible health effects of arsenic.

Arsenic has been shown to inhibit several enzymes, such as glutathione reductase (Styblo et al. 1997, Chouchane and Snow 2001, Muller et al. 1995), thioredoxin reductase (Lin et al. 1999, Lin et al. 2001), lipoamide dehydrogenase (Stevenson et al. 1978), and pyruvate dehydrogenase (Petrick et al. 2000). It may also affect proteins involved in signaling pathways (Bode and Dong 2002). Binding of arsenic with proteins via sulfhydryl groups is believed to be responsible for the observed effects of arsenic on these proteins. The present study provides direct evidence of arsenic-protein binding, using metallothionein binding with $As^{III}$, $MMA^{III}$ and $DMA^{III}$ as a model system. The concentrations of metallothionein (7 μM, or ~45 μg/mL) and arsenicals (as low as 0.35 μM) used in this study are relevant to those present in biological systems. The normal concentration of metallothionein in liver and kidney is on the order of tens of μg/g (Nordberg 1998). The concentrations of trivalent arsenicals in urine samples from humans exposed to elevated arsenic from drinking water are on the order of single digit μM (Aposhian et al. 2000a; Le et al. 2000a, 2000b; Del Razo et al. 2001; Mandal et al. 2001). The actual concentrations of trivalent arsenicals in the bladder could be higher because the trivalent arsenicals are readily oxidized in the urine (Gong et al. 2001).

Metallothionein can be induced by the arsenicals (Kreppel et al. 1993, Hochadel and Waalkes 1997, Flora and Tripathi 1998, Romach et al. 2000, Falnoga et al. 2000, Liu et al. 2001a). Since it was first recognized in equine renal cortex by Margoshes and Vallee in 1957 (Margoshes and Vallee 1957), it has been the subject of numerous studies (Klaassen and Liu 1998, Klaassen et al. 1999, Philcox et al. 1995, Dalton et al. 1996, Kojima et al. 1976, Dabrio et al. 2000, Mota et al. 2000) due to its important role in transport, free radical scavenging, and detoxification of heavy metals, especially cadmium, zinc, copper, and mercury, all of which induce its production. However, the mechanism of metallothionein induction by these cations and the arsenicals is unknown.

Arsenic reacts with protein sulfhydryl groups; thus, metallothionein could be a potentially protective protein. It has a high affinity for many metals and is known to effectively protect cells from cadmium toxicity. Liu et al. (2000) found that MT-null mice are more sensitive than wild-type mice to the hepatotoxic and nephrotoxic effects of chronic exposure to arsenic. Park et al. (2001) showed that the $LD_{50}$ of arsenic for wild-type mice was 1.4-fold higher than for MT-null mice. The present studies on the interaction of MT with arsenic suggest that the protective effect may be a result of binding of the arsenic species by MT.

Maitani et al. (1987) found that $As^{III}$ but not $As^{V}$, $MMA^{V}$, or $DMA^{V}$ was a modest inducer of MT in the liver of mice. $As^{III}$ did not appear to be bound to MT in appreciable amounts. Kreppel et al. (1994) reported that pretreatment of mice with zinc, an inducer of MT, protected against the acute toxicity of $As^{III}$. However, the protective effect of zinc pretreatment did not show a correlation with the extent of induction of MT. There was very little arsenic bound to MT in the cytosol of the Zn-pretreated mice, as determined by gel-filtration fractionation analysis. The exact role played by MT in the cellular response to arsenic is still unclear. There may be some secondary process related to the induction of MT that alters the metabolism or cellular response of arsenic that affects its toxicity.

The interaction of arsenic with MT may serve as a model for the interaction of arsenic with other cellular proteins, including DNA repair proteins. The effect of arsenic on DNA repair is a potential mechanism of action underlying carcinogenesis. The technique described here has the potential to be used in studies of the interaction of arsenic with these other proteins. We are currently studying the interaction of arsenicals with hemoglobin. One objective is to understand the observed difference between rodents and humans in the retention of arsenic in red blood cells. The technique also provides an approach to study arsenic effects on DNA repair proteins, which is a topic under continuing investigation.

## REFERENCES


Abbas, A.K., A.H. Lichtman, and J.S. Pober. 1994. *Cellular and Molecular Immunology*. 2nd ed. Philadelphia: W.B. Saunders Company.

Abernathy, C.O., W.R. Chappell, M.E. Meek, H. Gibb, and H.R. Guo. 1996. Is Ingested Inorganic Arsenic a "Threshold" Carcinogen? *Fundam. Appl. Toxicol.,* 29:168-175.

Abernathy, C.O., Y.P. Liu, D. Longfellow, H.V. Aposhian, B.D. Beck, B.A. Fowler, R.A. Goyer, R. Menzer, T. Rossman, C. Thompson, and M. Waalkes. 1999. Arsenic: Health Effects, Mechanisms of Actions, and Research Issues. *Environ. Health Perspect.,* 107:593-597.

Allalunis-Turner, M.J., P.K. Zia, G.M. Barron, R. Mirzayans, and R.S. Day. 1995. Radiation-Induced DNA Damage and Repair in Cells of a Radiosensitive Human Malignant Glioma Cell Line. *Radiat. Res.,* 144:288-293.

Andrews, C.L., P. Vouros, and A. Harsch. 1999. Analysis of DNA Adducts Using High-Performance Separation Techniques Coupled to Electrospray Ionization Mass Spectrometry. *J. Chromatogr. A,* 856:515-526.

Annan, R.S., M.J. Huddleston, R. Verma, R.J. Deshaies, and S.A. Carr. 2001. A Multidimensional Electrospray MS-Based Approach to Phosphopeptide Mapping. *Anal. Chem.,* 73(3):393-404.

Annan, R.S., G.M. Kresbach, R.W. Giese, and P. Vouros. 1989. Trace Detection of Modified DNA Bases via Moving-Belt Liquid-Chromatography Mass-Spectrometry Using Electrophoretic Derivatization and Negative Chemical Ionization. *J. Chromatogr.,* 465(2):285-296.

Aposhian, H.V. 1997. Enzymatic Methylation of Arsenic Species and Other new Approaches to Arsenic Toxicity. *Annu. Rev. Pharmacol. Toxicol.,* 37:397-419.

Aposhian, H.V., E.S. Gurzau, X.C. Le, A. Gurzau, S.M. Healy, X. Lu, M. Ma, L. Yip, R.A. Zakharyan, R.M. Maiorino, R.C. Dart, M.G. Tirus, D. Gonzalez-Ramirez, D.L. Morgan, D. Avram, and M.M. Aposhian. 2000a. Occurrence of Monomethylarsonous Acid ($MMA^{III}$) in Urine of Humans Exposed to Inorganic Arsenic. *Chem. Res. Toxicol.,* 13:693-697.

Aposhian, H.V., R.A. Zakharyan, E.K. Wildfang, S.M. Healy, J. Gailer, T.R. Radabaugh, G.M. Bogdan, L.A. Powell, and M.M. Aposhian. 1999. How is Inorganic Arsenic Detoxified? In *Arsenic Exposure and Health Effects*. Edited by Chappell, W.R., C.O. Abernathy, and R.L. Calderon. Amsterdam: Elsevier.

Aposhian, H.V., B. Zheng, M.M. Aposhian, X.C. Le, M.E. Cebrian, W.R. Cullen, R.A. Zakharyan, M. Ma, R.C. Dart, Z. Cheng, P. Andrews, L. Yip, G.F. O'Malley, R.M. Maiorino, W. Van Voorhies, S.M. Healy, and A. Titcomb. 2000b. DMPS-Arsenic Challenge Test: Modulation of Arsenic Species, Including Monomethylarsonous Acid, Excreted in Human Urine. *Toxicol. Appl. Pharmacol.,* 165:74-83.

Astolfi, E., A. Maccagno, J.C.G. Fernandez, R. Vaccaro, and R. Stimola. 1981. Relation Between Arsenic in Drinking Water and Skin Cancer. *Biol. Trace Elem. Res.,* 3(2):133-143.

ATDSR (Agency for Toxic Substances and Disease Registry). 2000. *Toxicological Profile for Arsenic*. Agency for Toxic Substances and Disease Registry.

Atta, M., M.E. Lafferty, M.K. Johnson, J. Gaillard, and J. Meyer. 1998. Heterologous Biosynthesis and Characterization of the [2Fe-2S]-containing N-terminal Domain of *Clostridium pasteurianum* Hydrogenase. *Biochemistry,* 37(45):15974-15980.

Baan, R.A., M.J.S.T. Steenwinkel, S. van Asten, R. Roggeband, and J.H.M. van Delft. 1997. The Use of Benzo(a)pyrene Diolepoxide-Modified DNA Standards for Adduct Quantification in $^{32}$P-postlabeling to Assess Exposure to Polycyclic Aromatic Hydrocarbons: Application in Biomonitoring Study. *Mutat. Res.,* 378:41-50.

Baan, R.A., P.T.M. van den Berg, W.P. Watson, and R.J. Smith. 1988. *In situ* Detection of DNA Adducts Formed in Cultured Cells by Benzo(a)pyrene Diolepoxide (BPDE), With Monoclonal Antibodies Specific for the BP-Deoxyguanosine Adduct. *Toxicol. Environ. Chem.,* 16:325-339.

Bagla, P., and J. Kaiser. 1996. India's Spreading Health Crisis Draws Global Arsenic Experts. *Science,* 274:174-175.

Bakkenist, C.J., and M.B. Kastan. 2003. DNA Damage Activates ATM Through Intermolecular Autophosphorylation and Dimer Dissociation. *Nature,* 421:499-506.

Banin, S., L. Moyal, S. Shieh, Y. Taya, C.W. Anderson, L. Chessa, N.I. Smorodinsky, C. Prives, Y. Reiss, Y. Shiloh, and Y. Ziv. 1998. Enhanced Phosphorylation of p53 by ATM in Response to DNA Damage. *Science,* 281:1674-1677.

Barry, J.P., C. Norwood, and P. Vouros. 1996. Detection and Identification of Benzo(a)pyrene Diol Epoxide Adducts to DNA Utilizing Capillary Electrophoresis-Electrospray Mass Spectrometry. *Anal. Chem.,* 68:1432-1438.

Basu, A., J. Mahata, S. Gupta, and A.K. Giri. 2001. Genetic Toxicology of a Paradoxical Human Carcinogen, Arsenic: A Review. *Mutat. Res.,* 488:171-194.

Batty, D.P., and R.D. Wood. 2000. Damage Recognition in Nucleotide Excision Repair of DNA. *Gene,* 241:193-204.

Beach, A.C., and R.C. Gupta. 1992. Human Biomonitoring and the $^{32}$P-Postlabeling Assay. *Carcinogenesis,* 13:1053-1074.

Beck, J.L., M.L. Colgrave, S.F. Ralph, and M.M. Sheil. 2001. Electrospray Ionization Mass Spectrometry of Oligonucleotide Complexes With Drugs, Metals, and Proteins. *Mass Spectrom. Rev.,* 20(2):61-87.

Beland, F.A., D.R. Doerge, M.I. Churchwell, M.C. Poirier, B. Schoket, and M.M. Marques. 1999. Synthesis, Characterization, and Quantitation of a 4-Aminobiphenyl-DNA Adduct Standard. *Chem. Res. Toxicol.,* 12:68-77.

Bernstam, L., and J. Nriagu. 2000. Molecular Aspects of Arsenic Stress. *J. Toxicol. Environ. Health B Crit. Rev.,* 3(4):293-322.

Bode, A.M., and Z.G. Dong. 2002. The Paradox of Arsenic: Molecular Mechanisms of Cell Transformation and Chemotherapeutic Effects. *Crit. Rev. Oncol. Hemat.,* 42:5-24.

Bogdan, G.M., A. Sampayo-Reyes, and H.V. Aposhian. 1994. Arsenic Binding Proteins of Mammalian Systems: I. Isolation of Three Arsenite-Binding Proteins of Rabbit Liver. *Toxicology,* 93:175-193.

Booth, E.D., J.P. Aston, P.T.M. van den Berg, R.A. Baan, D.A. Riddick, L.T. Wade, A.S. Wright, and W.P. Watson. 1994. Class-Specific Immunoadsorption Purification for Polycyclic Aromatic Hydrocarbon-DNA Adducts. *Carcinogenesis,* 15:2099-2106.

Bremner, I. 1991. Nutritional and Physiologic Significance of Metallothionein. *Methods Enzymol.,* 205:25-35.

Buchet, J.P., and R. Lauwerys. 1994. Inorganic Arsenic Metabolism in Humans. In *Arsenic Exposure and Health.* Edited by Chappell, W.R., C.O. Abernathy, and C.R. Cothern. Northwood, UK: Science and Technology Letters.

Buchet, J.P., R. Lauwerys, and H. Roels. 1981. Comparison of the Urinary Excretion of Arsenic Metabolites After a Single Dose of Sodium Arsenite, Monomethylarsonate or Dimethylarsinate in Man. *Int. Arch. Occup. Environ. Health,* 48:71-79.

Buchet, J.P., D. Lison, M. Ruggeri, V. Foa, and G. Elia. 1996. Assessment of Exposure to Inorganic Arsenic, a Human Carcinogen, due to the Consumption of Seafood. *Arch. Toxicol.,* 70(11):773-778.

Burrows, G.J., and E.E. Turner. 1920. A New Type of Compound Containing Arsenic. *J. Chem. Soc. Transact.,* 117:1373-1383.

Byron, W.R., G.W. Bierbower, J.B. Brouwer, and W.H. Hansen. 1967. Pathologic Changes in Rats and Dogs From Two-Year Feeding of Sodium Arsenite or Sodium Arsenate. *Toxicol. Appl. Pharmacol.,* 10:132-147.

Cadet, J., M. Berger, T. Douki, and J.L. Ravanat. 1997. Oxidative Damage to DNA: Formation, Measurement and Biological Significance. *Rev. Physiol. Biochem. Pharmacol.,* 131:1-87.

Cadet, J., F. Odin, J.F. Mouret, M. Polverelli, A. Audic, P. Giacomoni, A. Favier, and M.J. Richard. 1992. Chemical and Biochemical Postlabeling Methods for Singling Out Specific Oxidative DNA Lesions. *Mutat. Res.,* 275(3-6):343-354.

Cadet, J., and M. Weinfeld. 1993. Detecting DNA Damage. *Anal. Chem.,* 65:675A-682A.

Caldecott, K., and P. Jeggo. 1991. Cross-Sensitivity of Gamma-Ray-Sensitive Hamster Mutants to Cross-Linking Agents. *Mutat. Res.,* 255(2):111-121.

Caldecott, K.W. 2001. Mammalian DNA Single-Strand Break Repair: An X-Ra(y)ted Affair. *Bioessays,* 23:447-455.

Calderon, R.L., E. Hudgens, X.C. Le, D. Schreinemachers, and D.J. Thomas. 1999. Excretion of Arsenic in Urine as a Function of Exposure to Arsenic in Drinking Water. *Environ. Health Perspect.,* 107:663-667.

Canman, C.E., D.S. Lim, K.A. Cimprich, Y. Taya, K. Tamai, K. Sakaguchi, E. Appella, M.B. Kastan, and J.D. Siliciano. 1998. Activation of the ATM Kinase by Ionizing Radiation and Phosphorylation of p53. *Science,* 281:1677-1679.

Carlson-Lynch, H., B.D. Beck, and P.D. Boardman. 1994. Arsenic Risk Assessment. *Environ. Health Perspect.,* 102(4):354-356.

Carnelley, T.J., S. Barker, H. Wang, W.G. Tan, M. Weinfeld, and X.C. Le. 2001. Synthesis, Characterization, and Applications of a Fluorescent Probe of DNA Damage. *Chem. Res. Toxicol.,* 14(11):1513-1522.

Carney, J.P., R.S. Maser, H. Olivares, E.M. Davis, M. Le Beau, J.R. Yates, L. Hays, W.F. Morgan, and J.H. Petrini. 1998. The hMre11/hRad50 Protein Complex and Nijmegen Breakage Syndrome: Linkage of Double-Strand Break Repair to the Cellular DNA Damage Response. *Cell,* 93:477-486.

Cartwright, R., C.E. Tambini, P.J. Simpson, and J. Thacker. 1998. The XRCC2 DNA Repair Gene From Human and Mouse Encodes a Novel Member of the recA/RAD51 Family. *Nucleic Acids Res.,* 26(13):3084-3089.

Casale, G.P., M. Singhal, S. Bhattacharya, R. Ramanathan, K.P. Roberts, D.C. Barbacci, J. Zhao, R. Jankowiak, M.L. Gross, E.L. Cavalieri, G.J. Small, S.I. Rennard, J.L. Mumford, and M.L. Shen. 2001. Detection and Quantification of Depurinated Benzo(a)pyrene-Adducted DNA Bases in the Urine of Cigarette Smokers and Women Exposed to Household Coal Smoke. *Chem. Res. Toxicol.,* 14(2):192-201.

CDC (Center for Disease Control). 2001. *International Classification of Diseases, Ninth Revision (ICD - 9). Mortality Data From the National Vital Statistics System.* Online URL: www.cdc.gov/nchs/about/major/dvs/icd9des.html. Center for Disease Control.

Cebrian, M.E., A. Albores, M. Aguilar, and E. Blakely. 1983. Chronic Arsenic Poisoning in the North of Mexico. *Human Toxicol.,* 2:121-133.

Chappell, W.R., C.O. Abernathy, and R.L. Calderon, eds. 1999. *Arsenic Exposure and Health Effects.* Amsterdam: Elsevier.

Chappell, W.R., B.D. Beck, K.G. Brown, R. Chaney, C.R. Cothern, K.J. Irgolic, D.W. North, I. Thornton, and T.A. Tsongas. 1997. Inorganic Arsenic: A Need and an Opportunity to Improve Risk Assessment. *Environ. Health Perspect.,* 105:1060-1067.

Chassaigne, H., and R. Lobinski. 1998. Speciation of Metal Complexes With Biomolecules by Reversed-Phase HPLC With Ion-Spray and Inductively Coupled Plasma Mass Spectrometric Detection. *Fresenius J. Anal. Chem.,* 361(3):267-273.

Chatterjee, A., D. Dass, B.K. Mandal, T.R. Chowdhury, G. Samanta, and D. Chakraborti. 1995. Arsenic in Ground Water in Six Districts of West Bengal, India: The Biggest Arsenic Calamity in the World. Part 1. Arsenic Species in Drinking Water and Urine of the Affected People. *Analyst,* 120:643-650.

Chen, C.J., C.W. Chen, M.M. Wu, and T.L. Kuo. 1992. Cancer Potential in Liver, Lung, Bladder, and Kidney due to Ingested Inorganic Arsenic in Drinking Water. *Br. J. Cancer,* 66:888-892.

Chen, C.J., Y.C. Chuang, T.M. Lin, and H.Y. Wu. 1985. Malignant Neoplasms Among Residents of a Blackfoot Disease-Endemic Area in Taiwan: High-Arsenic Artesian Well Water and Cancers. *Cancer Res.,* 45:5895-5899.

Chen, C.J., Y.C. Chuang, S.L. You, T.M. Lin, and H.Y. Wu. 1986. A Retrospective Study on Malignant Neoplasms of Bladder, Lung and Liver in Blackfoot Disease Endemic Area in Taiwan. *Br. J. Cancer,* 53(3):399-405.

Chen, C.M., H.L. Mobley, and B.P. Rosen. 1985. Separate Resistances to Arsenate and Arsenite (Antimonate) Encoded by the Arsenical Resistance Operon of R Factor R773. *J. Bacteriol.,* 161(2):758-763.

Chen, F.T.A., and J.C. Sternberg. 1994. Characterization of Proteins by Capillary Electrophoresis in Fused-Silica Columns – Review on Serum-Protein Analysis and Application to Immunoassays. *Electrophoresis,* 15(1):13-21.

Chen, G.Q., X.G. Shi, W. Tang, S.M. Xiong, J. Zhu, X. Cai, Z.G. Han, J.H. Ni, G.Y. Shi, P.M. Jia, M.M. Liu, K.L. He, C. Niu, J. Ma, P. Zhang, T.D. Zhang, P. Paul, T. Naoe, K. Kitamura, W. Miller, S. Waxman, Z.Y. Wang, H. de The, S.J. Chen, and Z. Chen. 1997. Use of Arsenic Trioxide ($As_2O_3$) in the Treatment of Acute Promyelocytic Leukemia (APL): I. $As_2O_3$ Exerts Dose-Dependent Dual Effects on APL Cells. *Blood,* 89:3345-3353.

Chen, H., and C. Horvath. 1995. High-Speed High-Performance Liquid Chromatography of Peptides and Proteins. *J. Chromatogr. A,* 705:3-20.

Chen, Y.C., S.Y. Lin-Shiau, and J.K. Lin. 1998. Involvement of Reactive Oxygen Species and Caspase 3 Activation in Arsenite-Induced Apoptosis. *J. Cell. Physiol.,* 177:324-333.

Chiarelli, M., and J.O. Lay, Jr. 1992. Mass Spectrometry for the Analysis of Carcinogen-DNA Adducts in Humans. *Mass Spectrom. Rev.,* 11:447-493.

Chiou, H.Y., Y.M. Hsueh, K.F. Liaw, S.F. Horng, M.H. Chiang, Y.S. Pu, J.S. Lin, C.H. Huang, and C.J. Chen. 1995. Incidence of Internal Cancers and Ingested Inorganic Arsenic: A Seven-Year Follow-Up Study in Taiwan. *Cancer Res.,* 55:1296-1300.

Choi, D.J., D.J. Marino-Alessandri, N.E. Geactinov, and D.A. Scicchitano. 1994. Site-Specific Benzo(a)pyrene Diol Epoxide-DNA Adducts Inhibit Transcription Elongation by Bacteriophage T7 RNA Polymerase. *Biochemistry,* 33:780-787.

Chory, J., and J.D. Pollard, Jr. 1999. Separation of Small DNA Fragments by Conventional Gel Electrophoresis. In *Current Protocols in Molecular Biology*. Edited by Ausubel, F.M., R. Brent, R.E. Kingston, D.M. Moore, J.G. Seidman, J.A. Smith, and K. Struhl. New York: John Wiley & Sons.

Chouchane, S., and E.T. Snow. 2001. *In vitro* Effect of Arsenical Compounds Oil Glutathione-Related Enzymes. *Chem. Res. Toxicol.,* 14:517-522.

Chowdhury, T.R., B.K. Mandal, G. Samanta, G.K. Basu, P.P. Chowdhury, C.R. Chanda, N.K. Karan, D. Lodh, R.K. Dhar, D. Das, K.C. Saha, and D. Chakraborti. 1997. Arsenic in Ground Water in Six Districts of West Bengal, India: The Biggest Arsenic Calamity in the World. The Status Report up to August 1995. In *Arsenic Exposure and Health Effects*. Edited by Abernathy, C.O., R.L. Calderon, and W.R. Chappell. London: Chapman and Hall.

Cleaver, J.E., and K.H. Kraemer. 1989. Xeroderma Pigmentosum. In *The Metabolic Basis of Inherited Disease II*. 6th ed. Edited by Scriver, C.R., A.L. Beaudet, and W.S. Sly. New York: McGraw-Hill.

Concha, G., G. Vogler, D. Lezcano, B. Nermell, and M. Vahter. 1998. Exposure to Inorganic Arsenic Metabolites During Early Human Development. *Toxicol. Sci.,* 44:185-190.

Conney, A.H., R.L. Chang, D.M. Jerina, and S.J.C. Wei. 1994. Studies on the Metabolism of Benzo(a)pyrene and Dose-Dependent Differences in the Mutagenic Profile of its Ultimate Carcinogenic Metabolite. *Drug Metab. Rev.,* 26:125-163.

Cooper, P.K., T. Nouspikel, S.G. Clarkson, and S.A. Leadon. 1997. Defective Transcription-Coupled Repair of Oxidative Base Damage in Cockayne Syndrome Patients From XP Group G. *Science,* 275:990-993.

Coren, M.H., S. Hirschfeld, S. Flamm Honig, A. Ibrahim, J.R. Johnson, J.J. O'Leary, R.M. White, G.A. Williams, and R. Pazdur. 2001. Drug Approval Summaries: Arsenic Trioxide, Tamoxifen Citrate, Anastrazole, Paclitaxel, Bexarotene. *Oncologist,* 6:411-422.

Cosman, M., V. Ibanez, N.E. Geactinov, and R.G. Harvey. 1990. Preparation and Isolation of Adducts in High Yield Derived From the Binding of Two Benzo(a)pyrene-7,8-dihydroxy-9,10-oxide Stereoisomers to the Oligonucleotide d(ATATGTATA). *Carcinogenesis,* 11:1667-1672.

Crecelius, E.A. 1977. Changes in the Chemical Speciation of Arsenic Following Ingestion by Man. *Environ. Health Perspect.,* 19:147-150.

Crecelius, E.A., and J. Yager. 1997. Intercomparison of Analytical Methods for Arsenic Speciation in Human Urine. *Environ. Health Perspect.,* 105:650-653.

Cullen, W.R., B.C. McBride, H. Manji, A.W. Pickett, and J. Reglinski. 1989. The Metabolism of Methylarsine Oxide and Sulfide. *Appl. Organomet. Chem.,* 3:71-78.

Cullen, W.R., B.C. McBride, and J. Reglinski. 1984. The Reduction of Trimethylarsine Oxide to Trimethylarsine by Thiols: A Mechanistic Model for the Biological Reduction of Arsenicals. *J. Inorg. Biochem.,* 21:45-60 and 179-194.

Cullen, W.R., and K.J. Reimer. 1989. Arsenic Speciation in the Environment. *Chem. Rev.*, 89:713-764.

Dabrio, M., and A.R. Rodriguez. 2000. Study of Complexing Properties of the Alpha-Metallothionein Domain With Cadmium and/or Zinc, Using Differential Pulse Polarography. *Anal. Chim. Acta*, 424(1):77-90.

Dabrio, M., G. Van Vyncht, G. Bordin, and A.R. Rodriguez. 2001. Study of Complexing Properties of the Alpha and Beta Metallothionein Domains With Cadmium and/or Zinc Using Electrospray Ionisation Mass Spectrometry. *Anal. Chim. Acta*, 435(2):319-330.

Dabrio, M., V. Virtanen, G. Bordin, and A.R. Rodriguez. 2000. Contribution to the Study of Complexing Properties of Zn-Metallothionein by CZE-DAD. *Talanta*, 53(3):587-598.

Dalton, T., K. Fu, R.D. Palmiter, and G.K. Andrews. 1996. Transgenic Mice That Overexpress Metallothionein-I Resist Dietary Zinc Deficiency. *J. Nutr.*, 126(4):825-833.

Das, D., A. Chatterjee, B.K. Mandal, G. Samanta, D. Chakraborti, and B. Chanda. 1995. Arsenic in Ground-Water in 6 Districts of West-Bengal, India – the Biggest Arsenic Calamity in the World. Part 2. Arsenic Concentration in Drinking-Water, Hair, Nails, Urine, Skin-Scale and Liver-Tissue (Biopsy) of the Affected People. *Analyst*, 120(3):917-924.

Das, D., A. Chatterjee, G. Samanta, B. Mandal, T.R. Chowdhury, P.P. Chowdhury, C. Chanda, G. Basu, D. Lodh, S. Nandi, T. Chakraborty, S. Mandal, S.M. Bhattacharya, and D. Chakraborti. 1994. Arsenic Contamination in Groundwater in 6 Districts of West-Bengal, India – the Biggest Arsenic Calamity in the World. *Analyst*, 119(12):N168-N170.

Debenham, P.G., N.J. Jones, and M.B.T. Webb. 1988. Vector-Mediated DNA Double-Strand Break Repair Analysis in Normal, and Radiation-Sensitive, Chinese Hamster V79 Cells. *Mutat. Res.*, 199(1):1-9.

Deforce, D.L.D., F.P.K. Ryniers, E.G. Van den Eeckhout, F. Lemiere, and E.L. Esmans. 1996. Analysis of DNA Adducts in DNA Hydrolysates by Capillary Zone Electrophoresis and Capillary Zone Electrophoresis-Electrospray Mass Spectrometry. *Anal. Chem.*, 68(20):3575-3584.

Del Razo, L.M., G.G. Garcia-Vargas, H. Vargas, A. Albores, M.E. Gonsebatt, R. Montero, P. Ostrosky-Wegman, M. Kelsh, and M.E. Cebrian. 1997. Altered Profile of Urinary Arsenic Metabolites in Adults With Chronic Arsenicism. A Pilot Study. *Arch. Toxicol.*, 71:211-217.

Del Razo, L.M., M. Styblo, W.R. Cullen, and D.J. Thomas. 2001. Determination of Trivalent Methylated Arsenicals in Biological Matrices. *Toxicol. Appl. Pharmacol.*, 174(3):282-293.

Delnomdedieu, M., M.M. Basti, M. Styblo, J.D. Otvos, and D.J. Thomas. 1994. Complexation of Arsenic Species in Rabbit Erythrocytes. *Chem. Res. Toxicol.*, 7(5):621-627.

Denissenko, M.F., A. Pao, M. Tang, and G.P. Pfeifer. 1996. Preferential Formation of Benzo(a)pyrene Adducts at Lung Cancer Mutational Hotspots in *p53*. *Science*, 274:430-432.

Dickey, C., R.M. Santella, D. Hattis, D. Tang, Y. Hsu, T. Cooper, T.L. Young, and F.P. Perera. 1997. Variability in PAH-DNA Adduct Measurements in Peripheral Mononuclear Cells: Implications for Quantitative Cancer Risk Assessment. *Risk Anal.*, 17:649-656.

Ding, J.M., T. Barlow, A. Dipple, and P. Vouros. 1998. Separation and Identification of Positively Charged and Neutral Nucleoside Adducts by Capillary Electrochromatography Microelectrospray Mass Spectrometry. *J. Am. Soc. Mass Spectrom.*, 9(8):823-829.

Ding, J.M., and P. Vouros. 1997. Capillary Electrochromatography and Capillary Electrochromatography Mass Spectrometry for the Analysis of DNA Adduct Mixtures. *Anal. Chem.,* 69(3):379-384.

Dong, J.T., and X.M. Luo. 1993. Arsenic-Induced DNA-Strand Breaks Associated With DNA-Protein Crosslinks in Human Fetal Lung Fibroblasts. *Mutat. Res.,* 302:97-102.

Driggers, W.J., V.I. Grishko, S.P. LeDoux, and G.L. Wilson. 1996. Defective Repair of Oxidative Damage in the Mitochondrial DNA of a Xeroderma Pigmentosum Group A Cell Line. *Cancer Res.,* 56:1262-1266.

Duhachek, S.D., J.R. Kenseth, G.P. Casale, G.J. Small, M.D. Porter, and R. Jankowiak. 2000. Monoclonal Antibody-Gold Biosensor Chips for Detection of Depurinating Carcinogen-DNA Adducts by Fluorescence Line-Narrowing Spectroscopy. *Anal. Chem.,* 72(16):3709-3716.

Edman, L., U. Mets, and R. Rigler. 1996. Conformational Transitions Monitored for Single Molecules in Solution. *Proc. Natl. Acad. Sci. USA,* 93:6710-6715.

Eggeling, C., J.R. Fries, L. Brand, R. Gunther, and C.A.M. Seidel. 1998. Monitoring Conformational Dynamics of a Single Molecule by Selective Fluorescence Spectroscopy. *Proc. Natl. Acad. Sci. USA,* 95:1556-1561.

EPA (Environmental Protection Agency). 2001. *Drinking Water Standards for Arsenic.* Online URL: http://www.epa.gov/ogwdw000/ars/arsenic.finalrule.pdf. Environmental Protection Agency.

Evangelista, R.A., and F.T.A. Chen. 1994. Analysis of Structural Specificity in Antibody-Antigen Reactions by Capillary Electrophoresis With Laser-Induced Fluorescence Detection. *J. Chromatogr. A,* 680(2):587-591.

Fairbairn, D.W., P.L. Olive, and K.L. O'Neill. 1995. The Comet Assay: A Comprehensive Review. *Mutat. Res.,* 339(1):37-59.

Falnoga, I., E. Stibilj, M. Tusek-Znidaric, Z. Slejkovec, D. Mazej, R. Jacimovic, and J. Scancar. 2000. Effect of Arsenic Trioxide on Metallothionein and its Conversion to Different Arsenic Metabolites in Hen Liver. *Biol. Trace Elem. Res.,* 78(1-3):241-254.

Fang, A.H., W.A. Smith, P. Vouros, and R.C. Gupta. 2001. Identification and Characterization of a Novel Benzo(a)pyrene-Derived DNA Adduct. *Biochem. Biophys. Res. Commun.,* 281:383-389.

Farrer, B.T., C.P. McClure, J.E. Penner-Hahn, and V.L. Pecoraro. 2000. Arsenic(III)-Cysteine Interactions Stabilize Three-Helix Bundles in Aqueous Solution. *Inorg. Chem.,* 39(24):5422-5423.

Feldmann, J., V.W.M. Lai, W.R. Cullen, M. Ma, X. Lu, and X.C. Le. 1999. Sample Preparation and Storage Can Change Arsenic Speciation in Human Urine. *Clin. Chem.,* 45:1988-1997.

Flora, S.J., and N. Tripathi. 1998. Hepatic and Renal Metallothionein Induction Following Single Oral Administration of Gallium Arsenide in Rats. *Biochem. Mol. Biol. Int.,* 45(6):1121-1127.

Floyd, R.A., J.J. Watson, P.K. Wong, D.H. Altmiller, and R.C. Rickard. 1986. Hydroxyl Free Radical Adduct of Deoxyguanosine: Sensitive Detection and Mechanisms of Formation. *Free Radic. Res. Commun.,* 1(3):163-172.

Foà, V., A. Colombi, M. Maroni, M. Burrati, and G. Calzaferri. 1984. The Speciation of the Chemical Forms of Arsenic in the Biological Monitoring of Exposure to Inorganic Arsenic. *Sci. Total Environ.,* 34:241-259.

Freeman, H.C., J.F. Uthe, R.B. Fleming, P.H. Odense, R.G. Ackman, G. Landry, and C.J. Musial. 1979. Clearance of Arsenic Ingested by Man From Arsenic Contaminated Fish. *Bull. Environ. Contam. Toxicol.,* 22(1-2):224-229.

Friedberg, E.C., G.C. Walker, and W. Siede. 1995. *DNA Repair and Mutagenesis*. Washington, DC: ASM Press.

Fuciarelli, A.F., B.J. Wegher, E. Gajewski, M. Dizdaroglu, and W.F. Blakely. 1989. Quantitative Measurement of Radiation-Induced Base Products in DNA Using Gas Chromatography Mass-Spectrometry. *Radiat. Res.,* 119(2):219-231.

Fujiwara, Y. 1982. Defective Repair of Mitomycin C Crosslinks in Fanconi's Anemia and Loss in Confluent Normal Human and Xeroderma Pigmentosum Cells. *Biochim. Biophys. Acta,* 699(3):217-225.

Fuller, L.F., and R.B. Painter. 1988. A Chinese Hamster Ovary Cell-Line Hypersensitive to Ionizing Radiation and Deficient in Repair Replication. *Mutat. Res.,* 193(2):109-121.

Funk, M., I. Ponten, A. Seidel, and B. Jernstrom. 1997. Critical Parameters for Adduct Formation of the Carcinogen (+)-*anti*-benzo(a)pyrene-7,8-dihydrodiol-9,10-epoxide With Oligonucleotides. *Bioconjugate Chem.,* 8:310-317.

Galati, R., A. Zijno, R. Crebelli, G. Falasca, and F. Tomei. 2001. Detection of Antibodies to the Benzo(a)pyrene Diol Epoxide-DNA Adducts in Sera From Individuals Exposed to Low Doses of Polycyclic Aromatic Hydrocarbons. *J. Exp. Clin. Cancer Res.,* 20:351-364.

Gangl, E.T., R.J. Turesky, and P. Vouros. 1999. Determination of *in vitro*- and *in vivo*-Formed DNA Adducts of 2-amino-3-methylimidazo-4,5-f-quinoline by Capillary Liquid Chromatography-Microelectrospray Mass Spectrometry. *Chem. Res. Toxicol.,* 12(10):1019-1027.

Gatei, M., D. Young, K.M. Cerosaletti, A. Desai-Mehta, K. Spring, S. Kozlov, M.F. Lavin, R.A. Gatti, P. Concannon, and K. Khanna. 2000. ATM-Dependent Phosphorylation of Nibrin in Response to Radiation Exposure. *Nat. Genet.,* 25:115-119.

Gatti, R.A., A. Tward, and P. Concannon. 1999. Cancer Risk in ATM Heterozygotes: A Model of Phenotypic and Mechanistic Differences Between Missense and Truncating Mutations. *Mol. Genet. Metab.,* 68:419-423.

Gebel, T., P. Birkenkamp, S. Luthin, and H. Dunkelberg. 1998. Arsenic(III), but not Antimony(III), Induces DNA-Protein Crosslinks. *Anticancer Res.,* 18:4253-4257.

Gebel, T.W. 2001. Genotoxicity of Arsenical Compounds. *Int. J. Hyg. Environ. Health,* 203:249-262.

Gehrig, P.M., C. You, R. Dallinger, C. Gruber, M. Brouwer, J.H. Kagi, and P.E. Hunziker. 2000. Electrospray Ionization Mass Spectrometry of Zinc, Cadmium, and Copper Metallothioneins: Evidence for Metal-Binding Cooperativity. *Protein Sci.,* 9(2):395-402.

Giese, R.W. 1997. Detection of DNA Adducts by Electron Capture Mass Spectrometry. *Chem. Res. Toxicol.,* 10:255-270.

Goldman, R., L. Enewold, E. Pellizzari, J.B. Beach, E.D. Bowman, S.S. Krishnan, and P.G. Sheilds. 2001. Smoking Increases Carcinogenic Polycyclic Aromatic Hydrocarbons in Human Lung Tissue. *Cancer Res.,* 61:6367-6371.

Gong, Z., X. Lu, W.R. Cullen, and X.C. Le. 2001. Unstable Trivalent Arsenic Metabolites, Monomethylarsonous Acid and Dimethylarsinous Acid. *J. Anal. At. Spectrom.,* 16:1409-1413.

Gonsebatt, M.E., L. Vega, A.M. Salazar, R. Montero, P. Guzman, J. Blas, L.M. Del Razo, G.G. Garcia-Vargas, A. Albores, M.E. Cebrian, M. Kelsh, and P. Ostrosky-Wegman. 1997. Cytogenetic Effects in Human Exposure to Arsenic. *Mutat. Res.,* 386:219-228.

Gosling, J.P. 2000. *Immunoassays*. Oxford: Oxford University Press.

Govan, H.L., Y. Valles-Ayoub, and J. Braun. 1990. Fine-Mapping of DNA Damage and Repair in Specific Genomic Segments. *Nucleic Acids Res.,* 18(13):3823-3830.

Goyer, R.A. 1996. Toxic Effects of Metals. In *Casarett and Doull's Toxicology: The Basic Science of Poisons*. 5th ed. Edited by Klaassen, C.D. New York: McGraw-Hill.

Guo, X., H.M. Chan, R. Guevremont, and K.W. Siu. 1999. Analysis of Metallothioneins by Means of Capillary Electrophoresis Coupled to Electrospray Mass Spectrometry With Sheathless Interfacing. *Rapid Commun. Mass Spectrom.,* 13(6):500-507.

Gupta, R.C. 1996. $^{32}$P-Postlabeling for Detection of DNA Adducts. In *Technologies for Detection of DNA Damage and Mutations*. Edited by Pfeifer, G.P. New York: Plenum Press.

Hakala, E., and L. Pyy. 1995. Assessment of Exposure to Inorganic Arsenic by Determining the Arsenic Species Excreted in Urine. *Toxicol. Lett.,* 77:249-258.

Hakala, K., S. Auriola, A. Koivisto, and H. Lonnberg. 1999. Mass Spectrometric (HPLC/ESI-MS/MS) Quantification of Pyrimido-1,2-a-purin-10(3H)-1, a Guanine Adduct Formed by Reaction of Malondialdehyde With DNA. *J. Pharm. Biomed. Anal.,* 21(5):1053-1061.

Hanelt, S., R. Helbig, A. Hartmann, M. Lang, A. Seidel, and G. Speit. 1997. A Comparative Investigation of DNA Adducts, DNA Strand Breaks and Gene Mutations Induced by Benzo(a)pyrene and (±)-*anti*-benzo(a)pyrene-7,8-diol 9,10-oxide in Cultured Human Cells. *Mutat. Res.,* 390:179-188.

Harlow, E., and D. Lane. 1988. *Antibodies: A Laboratory Manual*. Cold Spring Harbor, NY: Cold Spring Harbor Laboratory Press.

Hartmann, A., and G. Speit. 1996. Effect of Arsenic and Cadmium on the Persistence of Mutagen-Induced DNA Lesions in Human Cells. *Environ. Mol. Mutagen,* 27(2):98-104.

Hartwig, A. 1998. Carcinogenicity of Metal Compounds: Possible Role of DNA Repair Inhibition. *Toxicol. Lett.,* 102-103:235-239.

Hartwig, A., U.D. Groblinghoff, D. Beyersmann, A.T. Natarajan, R. Filon, and L.H. Mullenders. 1997. Interaction of Arsenic(III) With Nucleotide Excision Repair in UV-Irradiated Human Fibroblasts. *Carcinogenesis,* 18:399-405.

Hathout, Y., D. Fabris, and C. Fenselau. 2001. Stoichiometry in Zinc Ion Transfer From Metallothionein to Zinc Finger Peptides. *Int. J. Mass Spectrom.,* 204(1-3):1-6.

Hecht, S.S. 1999. Tobacco Smoke Carcinogens and Lung Cancer. *J. Natl. Cancer Inst.,* 91(14):1194-1210.

Hei, T.K., S.X. Liu, and C. Waldren. 1998. Mutagenicity of Arsenic in Mammalian Cells: Role of Reactive Oxygen Species. *Proc. Natl. Acad. Sci. USA,* 95:8103-8107.

Hernandez, H., K.S. Hewitson, P. Roach, N.M. Shaw, J.E. Baldwin, and C.V. Robinson. 2001. Observation of the Iron-Sulfur Cluster in *Escherichia coli* Biotin Synthase by Nanoflow Electrospray Mass Spectrometry. *Anal. Chem.,* 73(17):4154-4161.

Hernandez-Zavala, A., L.M. Del Razo, C. Aguilar, G.G. Garcia-Vargas, V.H. Borja, A. Albores, and M.E. Cebrian. 1998. Alteration in Bilirubin Excretion in Individuals Chronically Exposed to Arsenic in Mexico. *Toxicol. Lett.,* 99(2):79-84.

Hertz-Picciotto, I., A.H. Smith, D. Holtzman, M. Lipsett, and G. Alexeeff. 1992. Synergism Between Occupational Arsenic Exposure and Smoking in the Induction of Lung Cancer [see comments]. *Epidemiology,* 3:23-31.

Hess, M.T., D. Gunz, N. Luneva, N.E. Geactinov, and H. Naegeli. 1997. Base Pair Conformation-Dependent Excision of Benzo(a)pyrene Diol Epoxide-Guanine Adducts by Human Nucleotide Excision Repair Enzymes. *Mol. Cell. Biol.,* 17:7069-7076.

Ho, I.C., and T.L. Lee. 2002. Arsenite Pretreatment Attenuates Benzo(a)pyrene Cytotoxicity in a Human Lung Adenocarcinoma Cell Line by Decreasing Cyclooxygenase-2 Levels. *J. Toxicol. Environ. Health A,* 65:245-263.

Hochadel, J.F., and M. Waalkes. 1997. Sequence of Exposure to Cadmium and Arsenic Determines the Extent of Toxic Effects in Male Fischer Rats. *Toxicology,* 116(1-3):89-98.

Hoeijmakers, J.H.J. 2001. Genome Maintenance Mechanisms for Preventing Cancer. *Nature,* 411:366-373.

Hopenhayn-Rich, C., M.L. Biggs, A. Fuchs, R. Bergoglio, E.E. Tello, H. Nicolli, and A.H. Smith. 1996a. Bladder Cancer Mortality Associated With Arsenic in Drinking Water in Argentina. *Epidemiology,* 7:117-124.

Hopenhayn-Rich, C., M.L. Biggs, and A.H. Smith. 1998. Lung and Kidney Cancer Mortality Associated With Arsenic in Drinking Water in Cordoba, Argentina. *Int. J. Epidemiol.,* 27:561-569.

Hopenhayn-Rich, C., M.L. Biggs, A.H. Smith, D.A. Kalman, and L.E. Moore. 1996b. Methylation Study of a Population Environmentally Exposed to Arsenic in Drinking Water. *Environ. Health Perspect.,* 104:620-628.

Hsu, I., M.C. Poirier, S.H. Yuspa, D. Grunberger, I.B. Weinstein, R.H. Yolken, and C.C. Harris. 1981. Measurement of Benzo(a)pyrene-DNA Adducts by Enzyme Immunoassays and Radioimmunoassay. *Cancer Res.,* 41:1090-1095.

Hsu, T.M., T.M. Liu, S. Amin, N.E. Geactinov, and R.M. Santella. 1995. Determination of Stereospecificity of Benzo(a)pyrene Diolepoxide-DNA Antisera With Site-Specifically Modified Oligonucleotides. *Carcinogenesis,* 16:2263-2265.

Hu, Y., L. Su, and E.T. Snow. 1998. Arsenic Toxicity is Enzyme Specific and its Effects on Ligation are not Caused by the Direct Inhibition of DNA Repair Enzymes. *Mutat. Res.,* 408(3):203-218.

Huang, H., A. Jemal, C. David, S.A. Barker, D.H. Swenson, and J.C. Means. 1998a. Analysis of DNA Adduct, S-2-(N-7-guanyl)ethyl-glutathione, by Liquid Chromatography Mass Spectrometry and Liquid Chromatography Tandem Mass Spectrometry. *Anal. Biochem.,* 265(1):139-150.

Huang, S.Y., C.S. Chang, J.L. Tang, H.F. Tien, T.L. Kuo, S.F. Huang, Y.T. Yao, W.C. Chou, C.Y. Chung, C.H. Wang, M.C. Shen, and Y.C. Chan. 1998b. Acute and Chronic Arsenic Poisoning Associated With Treatment of Acute Promyelocytic Leukemia. *Br. J. Haematol.,* 103:1092-1095.

Huff, J., P. Chan, and A. Nyska. 2000. Is the Human Carcinogen Arsenic Carcinogenic to Laboratory Animals? *Toxicol. Sci.,* 55:17-23.

Hughes, M.F. 2002. Arsenic Toxicity and Potential Mechanisms of Action. *Toxicol. Lett.,* 133(1):1-16.

IARC. 1972. Certain Polynuclear Aromatic Hydrocarbons and Heterocyclic Compounds. *IARC Monogr. Eval. Carcinog. Risks Hum.,* 3:45-268.

IARC. 1980. Arsenic and Arsenic Compounds. *IARC Monogr. Eval. Carcinog. Risks Hum.,* 23:39-141.

IARC. 1983. Polynuclear Aromatic Compounds. Part 1. Chemical, Environmental, and Experimental Data. *IARC Monogr. Eval. Carcinog. Risks Hum.,* 32:33-451.

IARC. 1987. Overall Evaluation of Carcinogenicity: An Updating of IARC Monographs Volumes 1 to 42. *IARC Monogr. Eval. Carcinog. Risks Hum.,* Supp 7.

Jankowiak, R., R.S. Cooper, D. Zamzow, G.J. Small, G. Doskocil, and A.M. Jeffrey. 1988. Fluorescence Line Narrowing Nonphotochemical Hole Burning Spectrometry: Femtomole Detection and High Selectivity for Intact DNA PAH Adducts. *Chem. Res. Toxicol.,* 1(1):60-68.

Jennerwein, M.M., and A. Eastman. 1991. A Polymerase Chain Reaction-Based Method to Detect Cisplatin Adducts in Specific Genes. *Nucleic Acids Res.,* 19(22):6209-6214.

Jennings, K.R. 2000. The Changing Impact of the Collision-Induced Decomposition of Ions on Mass Spectrometry. *Int. J. Mass Spectrom.,* 200(1-3):479-493.

Jensen, L.T., J.M. Peltier, and D.R. Winge. 1998. Identification of a Four Copper Folding Intermediate in Mammalian Copper Metallothionein by Electrospray Ionization Mass Spectrometry. *J. Biol. Inorg. Chem.,* 3(6):627-631.

Jha, A.N., M. Noditi, R. Nilsson, and A.T. Natarajan. 1992. Genotoxic Effects of Sodium Arsenite on Human Cells. *Mutat. Res.,* 284:215-221.

Ji, G., and S. Silver. 1992. Reduction of Arsenate to Arsenite by the ArsC Protein of the Arsenic Resistance Operon of *Staphylococcus aureus* Plasmid pI258. *Proc. Natl. Acad. Sci. USA,* 89:9474-9478.

Jiang, G., Z. Gong, X.-F. Li, W.R. Cullen, and X.C. Le. 2003a. Interaction of Trivalent Arsenicals With Metallothionein. Chem Res Toxicol., 16(7):873-80.

Jiang, G., X. Lu, Z. Gong, W.R. Cullen, and X.C. Le. 2003b. Trivalent Arsenic Species: Analysis, Stability, and Interaction With a Protein. In *Arsenic Exposure and Health Effects*, with permission from Elsevier. Edited by Chappell, W.R., C.O. Abernathy, and R.L. Calderon.

Jimenez, G.S., S.H. Khan, J.M. Stommel, and G.M. Wahl. 1999. p53 Regulation by Post-Translational Modification and Nuclear Retention in Response to Diverse Stress. *Oncogene,* 18:7656-7665.

Johnson, J.M., and C. Voegtlin. 1930. Arsenic Derivatives of Cysteine. *J. Biol. Chem.,* 89:27-31.

Johnson, R.D., N. Liu, and M. Jasin. 1999. Mammalian XRCC2 Promotes the Repair of DNA Double-Strand Breaks by Homologous Recombination. *Nature,* 401(6751):397-399.

Jones, J.A., and D.C. Liebler. 2000. Tandem MS Analysis of Model Peptide Adducts From Reactive Metabolites of the Hepatotoxin 1,1-dichloroethylene. *Chem. Res. Toxicol.,* 13(12):1302-1312.

Jones, N.J., R. Cox, and J. Thacker. 1987. Isolation and Cross-Sensitivity of X-Ray-Sensitive Mutants of V79-4 Hamster Cells. *Mutat. Res.,* 183(3):279-286.

Jorgensen, T.J., P.S. Russell, and D.A. McRae. 1995. Radioresistant DNA Synthesis in SV40-Immortalized Ataxia Telangiectasia Fibroblasts. *Radiat. Res.,* 143:219-223.

Jyonouchi, H., S. Sun, K. Iijima, M. Wang, and S.S. Hecht. 1999. Effects of *anti*-7,8-dihydroxy-9,10-tetrahydrobenzo(a)pyrene on Human Small Airway Epithelial Cells and the Protective Effects of Myo-inositol. *Carcinogenesis,* 20(1):139-145.

Kalinowski, D.P., S. Illenye, and B. Van Houten. 1992. Analysis of DNA Damage and Repair in Murine Leukemia-L1210 Cells Using a Quantitative Polymerase Chain Reaction Assay. *Nucleic Acids Res.,* 20(13):3485-3494.

Kalman, D.A., J. Hughes, G. van Belle, T. Burbacher, D. Bolgiano, K. Coble, N.K. Mottet, and L. Polissar. 1990. The Effect of Variable Environmental Arsenic Contamination on Urinary Concentrations of Arsenic Species. *Environ. Health Perspect.,* 89:145-151.

Kang, D.H., N. Rothman, M.C. Poirier, A. Greenberg, C.H. Hsu, B.S. Schwartz, M.E. Baser, J.D. Groopman, A. Weston, and P.T. Strickland. 1995. Interindividual Differences in the Concentration of 1-Hydroxypyrene-Glucuronide in Urine and Polycyclic Aromatic Hydrocarbon-DNA Adducts in Peripheral White Blood Cells After Charbroiled Beef Consumption. *Carcinogenesis,* 16:1079-1085.

Kanisawa, M., and H.A. Schroeder. 1969. Life Term Studies on the Effect of Trace Elements on Spontaneous Tumours in Mice and Rats. *Cancer Res.,* 29:892-895.

Karow, J.K., A. Constantinou, J.L. Li, S.C. West, and I.D. Hickson. 2000. The Bloom's Syndrome Gene Product Promotes Branch Migration of Holliday Junctions. *Proc. Natl. Acad. Sci. USA,* 97(6504-6508).

Kavanagh, P., M.E. Farago, I. Thornton, W. Goessler, D. Kuehnelt, C. Schlagenhaufen, and K.J. Irgolic. 1998. Urinary Arsenic Species in Devon and Cornwall Residents, UK. A Pilot Study. *Analyst,* 123:27-29.

Kelman, D.J., K.T. Lilga, and M. Sharma. 1988. Synthesis and Application of Fluorescent Labeled Nucleotides to Assay DNA Damage. *Chem. Biol. Interact.,* 66:85-100.

Kitchin, K.T. 2001. Recent Advances in Arsenic Carcinogenesis: Modes of Action, Animal Model Systems, and Methylated Arsenic Metabolites. *Toxicol. Appl. Pharmacol.,* 172(3):249-261.

Klaassen, C.D., and J. Liu. 1998. Induction of Metallothionein as an Adaptive Mechanism Affecting the Magnitude and Progression of Toxicological Injury. *Environ. Health Perspect.,* 106 Suppl 1:297-300.

Klaassen, C.D., J. Liu, and S. Choudhuri. 1999. Metallothionein: An Intracellular Protein to Protect Against Cadmium Toxicity. *Annu. Rev. Pharmacol. Toxicol.,* 39:267-294.

Klug, W.S., and M.R. Cummings. 1994. *Concepts of Genetics.* New York: MacMillan.

Klungland, A., M. Hoss, D. Gunz, A. Constantinou, S.G. Clarkson, P.W. Doetsch, P.H. Bolton, R.D. Wood, and T. Lindahl. 1999. Base Excision Repair of Oxidative DNA Damage Activated by XPG Protein. *Mol. Cell. Biol.,* 3:33-42.

Knudsen, C.B., I. Bjornsdottir, O. Jons, and S.H. Hansen. 1998. Detection of Metallothionein Isoforms From Three Different Species Using On-line Capillary Electrophoresis-Mass Spectrometry. *Anal. Biochem.,* 265(1):167-175.

Kojima, Y., C. Berger, B.L. Vallee, and J.H. Kagi. 1976. Amino-Acid Sequence of Equine Renal Metallotionein-1B. *Proc. Natl. Acad. Sci. USA,* 73(10):3413-3417.

Kreppel, H., J.W. Bauman, J. Liu, J.M. McKim, Jr., and C.D. Klaassen. 1993. Induction of Metallothionein by Arsenicals in Mice. *Fundam. Appl. Toxicol.,* 20(2):184-189.

Kreppel, H., J. Liu, Y. Liu, F.X. Reichl, and C.D. Klaassen. 1994. Zinc-Induced Arsenite Tolerance in Mice. *Fundam. Appl. Toxicol.,* 23:32-37.

Kuljukka-Rabb, T., K. Peltonen, S. Isotalo, S. Mikkonen, L. Rantanen, and K. Savela. 2001. Time and Dose-Dependent DNA Binding of PAHs Derived From Diesel Particle Extracts, Benzo(a)pyrene and 5-methylchrysene in a Human Mammary Carcinoma Cell Line (MCF-7). *Mutagenesis,* 16:353-358.

Kwong, Y.L., W.Y. Au, C.S. Chim, A. Pang, C. Suen, and R. Liang. 2001. Arsenic Trioxide- and Idarubicin-Induced Remissions in Relapsed Acute Promyelocytic Leukemia: Clinicopathological and Molecular Features of a Pilot Study. *Am. J. Hematol.,* 66:274-279.

Lackinger, D., W. Ruppitsch, M.H. Ramirez, M. Hirsch-Kauffmann, and M. Schweiger. 1998. Involvement of the Fanconi Anemia Protein FA-C in Repair Processes of Oxidative DNA Damages. *FEBS Lett.,* 440:103-106.

Lam, M.T., Q.H. Wan, C.A. Boulet, and X.C. Le. 1999. Competitive Immunoassay for Staphylococcal Enterotoxin-A Using Capillary Electrophoresis With Laser-Induced Fluorescence Detection. *J. Chromatogr. A,* 853:545-553.

Lausch, R., O.W. Reif, P. Riechel, and T. Scheper. 1995. Analysis of Immunoglobulin-G Using a Capillary Electrophoretic Affinity Assay With Protein-A and Laser-Induced Fluorescence Detection. *Electrophoresis,* 16(4):636-641.

Le Page, F., E.E. Kwoh, A. Avrutskaya, A. Gentil, S.A. Leadon, A. Sarasin, and P.K. Cooper. 2000. Transcription-Couple Repair of 8-oxoguanine: Requirement for XPG, TFIIH, and CSB and Implications for Cockayne Syndrome. *Cell,* 101:159-171.

Le, X.C., W.R. Cullen, and K.J. Reimer. 1993. Determination of Urinary Arsenic and Impact of Dietary Arsenic Intake. *Talanta,* 40:185-193.

Le, X.C., W.R. Cullen, and K.J. Reimer. 1994. Human Urinary Arsenic Excretion After One-Time Ingestion of Seaweed, Crab, and Shrimp. *Clin. Chem.,* 40:617-624.

Le, X.C., X. Lu, M. Ma, W.R. Cullen, H.V. Aposhian, and B. Zheng. 2000a. Speciation of Key Arsenic Metabolic Intermediates in Human Urine. *Anal. Chem.,* 72:5172-5177.

Le, X.C., and M. Ma. 1998. Short-Column Liquid Chromatography With Hydride Generation Atomic Fluorescence Detection for the Speciation of Arsenic. *Anal. Chem.,* 70:1926-1933.

Le, X.C., M. Ma, X. Lu, W.R. Cullen, H.V. Aposhian, and B. Zheng. 2000b. Determination of Monomethylarsonous Acid, a Key Arsenic Methylation Intermediate, in Human Urine. *Environ. Health Perspect.,* 108:1015-1018.

Le, X.C., M. Ma, and N.A. Wong. 1996. Speciation of Arsenic Compounds Using High-Performance Liquid Chromatography at Elevated Temperature and Selective Hydride Generation Atomic Fluorescence Detection. *Anal. Chem.,* 68:4501-4506.

Le, X.C., C. Scaman, Y. Zhang, J. Zhang, N.J. Dovichi, O. Hindsgaul, and M.M. Palcic. 1995. Analysis by Capillary Electrophoresis: Laser-Induced Fluorescence Detection of Oligosaccharides Produced From Enzyme Reactions. *J. Chromatogr. A,* 716:215-220.

Le, X.C., J.Z. Xing, J. Lee, S.A. Leadon, and M. Weinfeld. 1998. Inducible Repair of Thymine Glycol Detected by an Ultrasensitive Assay for DNA Damage. *Science,* 280:1066-1069.

Leadon, S.A. 1988. Immunological Probes for Lesions and Repair Patches in DNA. In *DNA Repair: A Laboratory Manual of Research Procedures*. Edited by Friedberg, E.C., and P.C. Hanawalt. New York: Dekker.

LeBlanc, J.C.Y. 1997. Use of Ionspray Mass Spectrometry in the Speciation and Elemental Characterization of Metallothioneins. *J. Anal. At. Spectrom.,* 12(5):525-530.

Lee, T.C., S.L. Kao, and L.H. Yih. 1991. Suppression of Sodium Arsenite-Potentiated Cytotoxicity of Ultraviolet Light by Cycloheximide in Chinese Hamster Ovary Cells. *Arch. Toxicol.,* 65:640-645.

Lee, T.C., N. Tanaka, P.W. Lamb, T.M. Gilmer, and J.C. Barrett. 1988. Induction of Gene Amplification by Arsenic. *Science,* 241:79-81.

Lee, T.C., S. Wang-Wuu, R.Y. Huang, K.C. Lee, and K.Y. Jan. 1986. Differential Effects of Pre- and Post-Treatment of Sodium Arsenite on the Genotoxicity of Methyl Methanesulfonate in Chinese Hamster Ovary Cells. *Cancer Res.,* 46:1854-1857.

Lee-Chen, S.F., J.R. Gurr, I.B. Lin, and K.Y. Jan. 1993. Arsenite Enhances DNA Double-Strand Breaks and Cell Killing of Methyl Methanesulfonate-Treated Cells by Inhibiting the Excision of Alkali-Labile Sites. *Mutat. Res.,* 294:21-28.

Lee-Chen, S.F., C.T. Yu, and K.Y. Jan. 1992. Effect of Arsenite on the DNA Repair of UV-Irradiated Chinese Hamster Ovary Cells. *Mutagenesis,* 7(1):51-55.

Lees-Miller, S.P., R. Godbout, D.W. Chan, M. Weinfeld, R.S. Day, G.M. Barron, and J. Allalunis-Turner. 1995. Absence of p350 Subunit of DNA-Activated Protein Kinase From a Radiosensitive Human Cell Line. *Science,* 267:1183-1185.

Lehmann, W.D., A. Schlosser, G. Erben, R. Pipkorn, D. Bossmeyer, and V. Kinzel. 2000. Analysis of Isoaspartate in Peptides by Electrospray Tandem Mass Spectrometry. *Protein Sci.,* 9(11):2260-2268.

Levine, L., E. Seaman, E. Hammerschlag, and H. Van Vunakis. 1966. Antibodies to Photoproducts of Deoxyribonucleic Acids Irradiated With Ultraviolet Light. *Science,* 153(744):1666-1667.

Lew, Y.S., A. Kolozsvary, S.L. Brown, and J.H. Kim. 2002. Synergistic Interaction With Arsenic Trioxide and Fractionated Radiation in Locally Advanced Murine Tumor. *Cancer Res.,* 62:4202-4205.

Li, D., K. Morimoto, T. Takeshita, and Y. Lu. 2001a. Formamidopyrimidine-DNA Glycosylase Enhances Arsenic-Induced DNA Strand Breaks in PHA-Stimulated and Unstimulated Human Lymphocytes. *Environ. Health Perspect.,* 109:523-526.

Li, D., F.F. Pervez, P.F. Firozi, L.E. Wang, C.H. Bosken, M.R. Spitz, W.K. Hong, and Q. Wei. 2001b. Sensitivity to DNA Damage Induced by Benzo(a)pyrene Diol Epoxide and Risk of Lung Cancer: A Case-Control Analysis. *Cancer Res.,* 61:1445-1450.

Li, D., M. Wang, L. Cheng, M.R. Spitz, W.N. Hittelman, and Q. Wei. 1996. *In vitro* Induction of Benzo(a)pyrene Diol Epoxide-DNA Adducts in Peripheral Lymphocytes as a Susceptibility Marker for Human Lung Cancer. *Cancer Res.,* 56:3638-3641.

Li, J., and C.M. Pickart. 1995. Binding of Phenylarsenoxide to Arg-tRNA Protein Transferase is Independent of Vicinal Thiols. *Biochemistry,* 34:15829-15837.

Li, M., R.J. Hurtubise, and A. Weston. 1999. Determination of Benzo(a)pyrene-DNA Adducts by Solid-Matrix Phosphorescence. *Anal. Chem.,* 71:4679-4683.

Lieber, M.R. 1999. The Biochemistry and Biological Significance of Nonhomologous DNA End Joining: An Essential Repair Process in Multicellular Eukaryotes. *Genes Cells,* 4:77-85.

Lim, C.K., Z.X. Yuan, R.M. Jones, I.N.H. White, and L.L. Smith. 1997. Identification and Mechanism of Formation of Potentially Genotoxic Metabolites of Tamoxifen: Study by LC-MS/MS. *J. Pharm. Biomed. Anal.,* 15(9-10):1335-1342.

Lim, D.S., S.T. Kim, B. Xu, R.S. Maser, J.H. Lin, J.H. Petrini, and M.B. Kastan. 2000. ATM Phosphorylates p95/nbs1 in an S-phase Checkpoint Pathway. *Nature,* 404:613-617.

Lin, S., W.R. Cullen, and D.J. Thomas. 1999. Methylarsenicals and Arsinothiols are Potent Inhibitors of Mouse Liver Thioredoxin Reductase. *Chem. Res. Toxicol.,* 12:924-930.

Lin, S., L.M. Del Razo, M. Styblo, C.Q. Wang, W.R. Cullen, and D.J. Thomas. 2001. Arsenicals Inhibit Thioredoxin Reductase in Cultured Rat Hepatocytes. *Chem. Res. Toxicol.,* 14:305-311.

Lin, S., Q. Shi, F.B. Nix, M. Styblo, M.A. Beck, K.M. Herbin-Davis, L.L. Hall, J.B. Simeonsson, and D.J. Thomas. 2002. A Novel S-Adenosyl-L-Methionine:Arsenic(III) Methyltransferase From Rat Liver Cytosol. *J. Biol. Chem.,* 277:10795-10803.

Lin, T.H., and Y.L. Huang. 1995. Chemical Speciation of Arsenic in Urine of Patients With Blackfoot Disease. *Biol. Trace Elem. Res.,* 48:251-261.

Lindahl, T., P. Karran, and R.D. Wood. 1997. DNA Excision Repair Pathways. *Curr. Opin. Genet. Dev.,* 7:158-169.

Liu, J., M.B. Kadiiska, Y. Liu, T. Lu, W. Qu, and M.P. Waalkes. 2001a. Stress-Related Gene Expression in Mice Treated With Inorganic Arsenicals. *Toxicol. Sci.,* 61(2):314-320.

Liu, J., Y. Liu, R.A. Goyer, W. Achanzar, and M.P. Waalkes. 2000. Metallothionein-I/II Null Mice are More Sensitive Than Wild-Type Mice to the Hepatotoxic and Nephrotoxic Effects of Chronic Oral or Injected Inorganic Arsenicals. *Toxicol. Sci.,* 55:460-467.

Liu, N., J.E. Lamerdin, R.S. Tebbs, D. Schild, J.D. Tucker, M.R. Shen, K.W. Brookman, M.J. Siciliano, C.A. Walter, W. Fan, L.S. Narayana, Z.Q. Zhou, A.W. Adamson, K.J. Sorensen, D.J. Chen, N.J. Jones, and L.H. Thompson. 1998. XRCC2 and XRCC3, New Human Rad51-Family Members, Promote Chromosome Stability and Protect Against DNA Cross-Links and Other Damages. *Mol. Cell,* 1(6):783-793.

Liu, S.X., M. Athar, I. Lippai, C. Waldren, and T.K. Hei. 2001b. Induction of Oxyradicals by Arsenic: Implication for Mechanism of Genotoxicity. *Proc. Natl. Acad. Sci. USA,* 98:1643-1648.

Lloyd, D.R., and P.C. Hanawalt. 2000. p53-Dependent Global Genomic Repair of Benzo(a)pyrene-7,8-diol-9,10-epoxide Adducts in Human Cells. *Cancer Res.,* 60:517-521.

Longo, J.A., B. Nevaldine, S.L. Longo, J.A. Winfield, and P.J. Hahn. 1997. An Assay for Quantifying DNA Double-Strand Break Repair That is Suitable for Small Numbers of Unlabeled Cells. *Radiat. Res.,* 147:35-40.

Loo, J.A. 1997. Studying Noncovalent Protein Complexes by Electrospray Ionization Mass Spectrometry. *Mass Spectrom. Rev.,* 16(1):1-23.

Lu, X., Z. Gong, M. Ma, W.R. Cullen, H.V. Aposhian, B. Zheng, E.S. Gurzau, and X.C. Le. 2001. Speciation of Human Urinary Arsenic Including the Trivalent Methylation Metabolites. In *Arsenic Exposure and Health Effects.* Edited by Chappell, W.R., C.O. Abernathy, and R.L. Calderon. New York: Elsevier Science.

Luo, Z.D., Y.M. Zhang, L. Ma, G.Y. Zhang, X. He, R. Wilson, D.M. Byrd, J.G. Griffiths, S. Lai, L. He, K. Grumski, and S.H. Lamm. 1997. Chronic Arsenicism and Cancer in Inner Mongolia – Consequences of Well-Water Arsenic Levels Greater Than 50 μg/L. In *Arsenic Exposure and Health Effects.* Edited by Abernathy, C.O., R.L. Calderon, and W.R. Chappell. London: Chapman and Hall.

Lynn, S., J.R. Gurr, H.T. Lai, and K.Y. Jan. 2000. NADH Oxidase Activation is Involved in Arsenite-Induced Oxidative DNA Damage in Human Vascular Smooth Muscle Cells. *Circ. Res.,* 86:514-519.

Lynn, S., H.T. Lai, J.R. Gurr, and K.Y. Jan. 1997. Arsenite Retards DNA Break Rejoining by Inhibiting DNA Ligation. *Mutagenesis,* 12:353-358.

Ma, M., and X.C. Le. 1998. Effect of Arsenosugar Ingestion on Urinary Arsenic Speciation. *Clin. Chem.,* 44:539-550.

Maier, A., B.L. Schumann, X. Chang, G. Talaska, and A. Puga. 2002. Arsenic Co-exposure Potentiates Benzo(a)pyrene Genotoxicity. *Mutat. Res.,* 517:101-111.

Maitani, T., N. Saito, A. Mayumi, S. Uchiyama, and Y. Saito. 1987. Chemical Form-Dependent Induction of Hepatic Zincthionein by Arsenic Administration and Effect of Co-Administered Selenium in Mice. *Toxicol. Lett.,* 39:63-70.

Mandal, B.K., Y. Ogra, and K.T. Suzuki. 2001. Identification of Dimethylarsinous and Monomethylarsonous Acids in Human Urine of the Arsenic-Affected Areas in West Bengal, India. *Chem. Res. Toxicol.,* 14:371-378.

Marafante, M., J. Rade, and E. Sabbioni. 1981. Intracellular Interaction and Metabolic Fate of Arsenite in the Rabbit. *Clin. Toxicol.,* 18:1335-1341.

Margoshes, M., and B.L. Vallee. 1957. A Cadmium Protein From Equine Kidney Cortex [Letter]. *J. Am. Chem. Soc.,* 79:4813-4814.

Margulis, L.A., V. Ibanez, and N.E. Geactinov. 1993. Base-Sequence Dependence of Covalent Binding of Benzo(a)pyrene Diol Epoxide to Guanine in Oligodeoxyribonucleotides. *Chem. Res. Toxicol.,* 6:59-63.

Mass, M.J., A. Tennant, R.C. Roop, W.R. Cullen, M. Styblo, D.J. Thomas, and A.D. Kligerman. 2001. Methylated Trivalent Arsenic Species are Genotoxic. *Chem. Res. Toxicol.,* 14(4):355-361.

Mass, M.J., and L. Wang. 1997. Arsenic Alters Cytosine Methylation Patterns of the Promoter of the Tumor Suppressor Gene p53 in Human Lung Cells: A Model for a Mechanism of Carcinogenesis. *Mutat. Res.,* 386:263-277.

Matsui, M., C. Nishigori, S. Toyokuni, J. Takada, M. Akaboshi, M. Ishikawa, S. Imamura, and Y. Miyachi. 1999. The Role of Oxidative DNA Damage in Human Arsenic Carcinogenesis: Detection of 8-hydroxy-2'-deoxyguanosine in Arsenic-Related Bowen's Disease. *J. Invest. Dermatol.,* 113:26-31.

McCloskey, J.A., and P.F. Crain. 1992. Progress in Mass Spectrometry of Nucleic Acid Constituents: Analysis of Xenobiotic Modifications and Measurements at High Mass. *Int. J. Mass Spectrom. Ion Process.,* 118:593-615.

Mei, N., N. Kunugita, T. Hirano, and H. Kasai. 2002. Acute Arsenite-Induced 8-hydroxyguanine is Associated With Inhibition of Repair Activity in Cultured Human Cells. *Biochem. Biophys. Res. Commun.,* 297:924-930.

Mei, N., J.Z. Xing, J. Hanson, X.C. Le, and M. Weinfeld. 2003. Genetic Predisposition to the Cytotoxicity of Arsenic. FASEB J., 17(15):2310-2312.

Melamede, R.J., Y.W. Kow, I.A. Bespalov, and S.S. Wallace. 1996. Detection of Oxidative DNA Base Damages: Immunochemical and Electrochemical Approaches. In *Technologies for Detection of DNA Damage and Mutations*. Edited by Pfeifer, G.P. New York: Plenum Press.

Melendez-Colon, V.J., A. Luch, A. Seidel, and W.M. Baird. 2000. Formation of Stable DNA Adducts and Apurinic Sites Upon Metabolic Activation of Bay and Fjord Region Polycyclic Aromatic Hydrocarbons in Human Cell Cultures. *Chem. Res. Toxicol.,* 13(1):10-17.

Menendez, D., G. Mora, A.M. Salazar, and P. Ostrosky-Wegman. 2001. ATM Status Confers Sensitivity to Arsenic Cytotoxic Effects. *Mutagenesis,* 16:443-448.

Merijanian, A., and R. Zingaro. 1966. Arsine Oxides. *Inorg. Chem.,* 5:187-191.

Meyer, J., J. Gagnon, L.C. Sieker, A. Van Dorsselaer, and J.M. Moulis. 1990. Rubredoxin From *Clostridium thermosaccharolyticum*: Amino Acid Sequence, Mass-spectrometric and Preliminary Crystallographic Data. *Biochem. J.,* 271(3):839-841.

Milton, A.H., Z. Hasan, and M. Rahman. 2001. Chronic Arsenic Poisoning and Respiratory Effects in Bangladesh. *J. Occup. Health,* 43(3):136-140.

Mitchell, D.L. 1996. Radioimmunoassay of DNA Damaged by Ultraviolet Light. In *Technologies for Detection of DNA Damage and Mutations*. Edited by Pfeifer, G.P. New York: Plenum Press.

Modrich, P. 1994. Mismatch Repair, Genetic Stability, and Cancer. *Science,* 266:1959-1960.

Modrich, P., and R. Lahue. 1996. Mismatch Repair in Replication Fidelity, Genetic Recombination, and Cancer Biology. *Annu. Rev. Biochem.,* 65:101-133.

Mota, J.P.V., A.R.L. Arias, M.R. Fernandez de la Campa, J.I.G. Alonso, and A. Sanz-Medel. 2000. Development of a Stable Isotope Approach for the Inductively Coupled Plasma-Mass Spectrometry Determination of Oxidized Metallothionein in Biological Materials. Anal. Biochem., 282(2):194-199.

Muller, M., F.J. Belas, I.A. Blair, and F.P. Guengerich. 1997. Analysis of 1,N-2-ethenoguanine and 5,6,7,9-tetrahydro-7- hydroxy-9-oxoimidazo 1,2-a purine in DNA Treated With 2-chlorooxirane by High Performance Liquid Chromatography Electrospray Mass Spectrometry and Comparison of Amounts to Other DNA Adducts. *Chem. Res. Toxicol.,* 10(2):242-247.

Muller, S., R.D. Walter, and A.H. Fairlamb. 1995. Differential Susceptibility of Filarial and Human Erythrocyte Glutathione-Reductase to Inhibition by the Trivalent Organic Arsenical Melarsen Oxide. *Mol. Biochem. Parasitol.,* 71:211-219.

Murai, T., H. Iwata, T. Otoshi, G. Endo, S. Horiguchi, and S. Fukushima. 1993. Renal Lesions Induced in F344 Ducrj Rats by 4-Weeks Oral Administration of Dimethylarsinic Acid. *Toxicol. Lett.,* 66(1):53-61.

Murphy, P. 2003. The Effects of BPDE and Arsenite on Genomic A549 DNA. Master's thesis. University of Alberta, Edmonton, Alta.

Murray, D., and E. Rosenberg. 1996. The Importance of the ERCC1/ERCC4[XPF] Complex for Hypoxic-Cell Radioresistance Does not Appear to Derive From its Participation in the Nucleotide Excision Repair Pathway. *Mutat. Res.,* 364:217-226.

Mushak, P., and A.F. Crocetti. 1995. Risk and Revisionism in Arsenic Cancer Risk Assessment. *Environ. Health Perspect.,* 103:684-689.

Nackerdien, Z.E., and D. Atha. 1996. Measurement of $^{60}$Co-ray Induced DNA Damage by Capillary Electrophoresis. *J. Chromatogr. B,* 683:85-89.

Nakagawa, K., Y. Taya, K. Tamai, and M. Yamaizumi. 1999. Requirement of ATM in Phosphorylation of the Human p53 Protein at Serine 15 Following DNA Double-Strand Breaks. *Mol. Cell. Biol.,* 19:2828-2834.

Natsume, T., H. Nakayama, O. Jansson, T. Isobe, K. Takio, and K. Mikoshiba. 2000. Combination of Biomolecular Interaction Analysis and Mass Spectrometric Amino Acid Sequencing. *Anal. Chem.,* 72(17):4193-4198.

Ng, J.C., D. Johnson, P. Imray, B. Chiswell, and M.R. Moore. 1998. Speciation of Arsenic Metabolites in the Urine of Occupational Workers and Experimental Rats Using an Optimised Hydride Cold-Trapping Method. *Analyst,* 123:929-933.

Nickson, R., J. McArthur, W. Burgess, K.M. Ahmed, P. Ravenscroft, and M. Rahman. 1998. Arsenic Poisoning of Bangladesh Groundwater. *Nature,* 395:338.

Nicotera, T.M. 1991. Molecular and Biochemical Aspects of Bloom's Syndrome. *Cancer Genet. Cytogenet.,* 53:1-13.

NIEHS (National Institute of Environmental Health Sciences). 1999. *Report on Carcinogens*. National Institute of Environmental Health Sciences.

Niu, S., S. Cao, and E. Shen. 1997. The Status of Arsenic Poisoning in China. In *Arsenic Exposure and Health Effects*. Edited by Chappell, W.R., C.O. Abernathy, and R.L. Calderon. Amsterdam: Elsevier.

Nordberg, M. 1998. Metallothioneins: Historical Review and State of Knowledge. *Talanta,* 43:243-254.

NRC (National Research Council). 1999. *Arsenic in Drinking Water*. Washington, DC: National Research Council.

NRC (National Research Council). 2001. *Arsenic in Drinking Water, Update*. Washington, DC: National Research Council.

Ochi, T., F. Nakajima, and N. Fukumori. 1998. Different Effects of Inorganic and Dimethylated Arsenic Compounds on Cell Morphology, Cytoskeletal Organization, and DNA Synthesis in Cultured Chinese Hamster V79 Cells. *Arch. Toxicol.,* 72:566-573.

Okui, T., and Y. Fujiwara. 1986. Inhibition of Human Excision DNA Repair by Inorganic Arsenic and the Co-mutagenic Effect in V79 Chinese Hamster Cells. *Mutat. Res.,* 172:69-76.

Painter, R.B. 1981. Radioresistant DNA Synthesis: An Intrinsic Feature of Ataxia Telangiectasia. *Mutat. Res.,* 84:183-190.

Papayannopoulos, I.A. 1995. The Interpretation of Collision-Induced Dissociation Tandem Mass-Spectra of Peptides. *Mass Spectrom. Rev.,* 14(1):49-73.

Park, J.D., Y.P. Liu, and C.D. Klaassen. 2001. Protective Effect of Metallothionein Against the Toxicity of Cadmium and Other Metals. *Toxicology,* 163(2-3):93-100.

Paull, T.T., E.P. Rogakou, V. Yamazaki, C.U. Kirchgessner, M. Gellert, and W.M. Bonner. 2000. A Critical Role for Histone H2AX in Recruitment of Repair Factors to Nuclear Foci After DNA Damage. *Curr. Biol.,* 10(15):886-895.

Pavanello, S., D. Favretto, F. Brugnone, G. Mastrangelo, G. Dal Pra, and E. Clonfero. 1999. HPLC/Fluorescence Determination of *anti*-BPDE-DNA Adducts in Mononuclear White Blood Cells From PAH-Exposed Humans. *Carcinogenesis,* 20:431-435.

Perera, F.P., M.C. Poirier, S.H. Yuspa, J. Nakayama, A. Jaretzki, M.M. Curnen, D.M. Knowles, and I.B. Weinstein. 1982. A Pilot Project in Molecular Cancer Epidemiology: Determination of Benzo(a)pyrene-DNA Adducts in Animal and Human Tissues by Immunoassays. *Carcinogenesis,* 3(12):1405-1410.

Pershagen, G., G. Nordberg, and N.E. Bjorklund. 1984. Carcinomas of the Respiratory Tract in Hamsters Given Arsenic Trioxide and/or Benzo(a)pyrene by the Pulmonary Route. *Environ. Res.,* 34:227-241.

Petillot, Y., E. Forest, J. Meyer, and J.M. Moulis. 1995. Observation of Holoprotein Molecular Ions of Several Ferredoxins by Electrospray-Ionization-Mass Spectrometry. *Anal. Biochem.,* 228(1):56-63.

Petrick, J.S., F. Ayala-Fierro, W.R. Cullen, D.E. Carter, and H.V. Aposhian. 2000. Monomethylarsonous Acid ($MMA^{III}$) is More Toxic Than Arsenite in Chang Human Hepatocytes. *Toxicol. Appl. Pharmacol.,* 163:203-207.

Petrick, J.S., J. Bhumasamudram, E.A. Mash, and H.V. Aposhian. 2001. Monomethylarsonous Acid (MMA III) and Arsenite: $LD_{50}$ in Hamsters and *in vitro* Inhibition of Pyruvate Dehydrogenase. *Chem. Res. Toxicol.,* 14:651-656.

Pfeifer, G.P., ed. 1996. *Technologies for Detection of DNA Damage and Mutations*. New York: Plenum Press.

Pfeifer, G.P., M.F. Denissenko, and M. Tang. 1998. PCR-Based Approaches to Adduct Analysis. *Toxicol. Lett.,* 102/103:447-451.

Pfeifer, G.P., R. Drouin, A.D. Riggs, and G.P. Holmquist. 1991. *In vivo* Mapping of a DNA Adduct at Nucleotide Resolution-Detection of Pyrimidine (6-4) Pyrimidone Photoproducts by Ligation-Mediated Polymerase Chain Reaction. *Proc. Natl. Acad. Sci. USA,* 88(4):1374-1378.

Philcox, J.C., P. Coyle, A. Michalska, K.H. Choo, and A.M. Rofe. 1995. Endotoxin-Induced Inflammation Does not Cause Hepatic Zinc Accumulation in Mice Lacking Metallothionein Gene Expression. *Biochem. J.,* 308:543-546.

Phillips, D.H. 1997. Detection of DNA Modifications by the $^{32}$P-postlabeling Assay. *Mutat. Res.,* 378(1-2):1-12.

Phillips, D.H., and M. Castegnaro. 1993. Results of an Interlaboratory Trial of $^{32}$P-postlabeling. *IARC Sci. Publ.,* 124:35-49.

Pi, J., H. Yamauchi, Y. Kumagai, G. Sun, T. Yashida, H. Aikawa, C. Hopenhayn-Rich, and N. Shimojo. 2002. Evidence for Induction of Oxidative Stress Caused by Chronic Exposure of Chinese Residents to Arsenic Contained in Drinking Water. *Environ. Health Perspect.,* 110:331-335.

Poirier, M.C. 1981. Antibodies to Carcinogen-DNA Adducts. *J. Natl. Cancer Inst.,* 67(3):515-519.

Poirier, M.C., R.M. Santella, I.B. Weinstein, D. Grunberger, and S.H. Yuspa. 1980. Quantitation of Benzo(a)pyrene-Deoxyguanosine Adducts by Radioimmunoassay. *Cancer Res.,* 40:412-416.

Pomroy, C., S.M. Charbonneau, R.S. McCullough, and G.K.H. Tam. 1980. Human Retention Studies With $^{74}$As. *Toxicol. Appl. Pharmacol.,* 53:550-556.

Pontius, F.W., K.G. Brown, and C.J. Chen. 1994. Health Implications of Arsenic in Drinking Water. *Jour. AWWA,* 86:52-63.

Pott, W.A., S.A. Benjamin, and R.S.H. Yang. 2001. Pharmacokinetics, Metabolism, and Carcinogenicity of Arsenic. *Rev. Environ. Contam. Toxicol.,* 169:165-214.

Rahman, M., M. Tondel, S.A. Ahmad, I.A. Chowdhury, M.H. Faruqee, and O. Axelson. 1999. Hypertension and Arsenic Exposure in Bangladesh. *Hypertension,* 33(1):74-78.

Randerath, K., and E. Randerath. 1994. P-32 Postlabeling Methods for DNA Adduct Detection – Overview and Critical Evaluation. *Drug Metab. Rev.,* 26(1-2):67-85.

Randerath, K., M.V. Reddy, and R.C. Gupta. 1981. P-32-Labeling Test for DNA Damage. *Proc. Natl. Acad. Sci. USA,* 78(10):6126-6129.

Roberts, D.W., M.I. Churchwell, F.A. Beland, J.L. Fang, and D.R. Doerge. 2001. Quantitative Analysis of Etheno-2'-deoxycytidine DNA Adducts Using On-line Immunoaffinity Chromatography Coupled With LC/ES-MS/MS Detection. *Anal. Chem.,* 73(2):303-309.

Rodin, S.N., and A.S. Rodin. 2000. Human Lung Cancer and p53: The Interplay Between Mutagenesis and Selection. *Proc. Natl. Acad. Sci. USA,* 97(22):12244-12249.

Rogakou, E.P., D.R. Pilch, A.H. Orr, V.S. Ivanova, and W.M. Bonner. 1998. DNA Double-Stranded Breaks Induce Histone H2AX Phosphorylation on Serine 139. *J. Biol. Chem.,* 273:5858-5868.

Rogers, K.R., A. Apostol, S.J. Madsen, and C.W. Spencer. 1999. Detection of Low Dose Radiation Induced DNA Damage Using Temperature Differential Fluorescence Assay. *Anal. Chem.,* 71(19):4423-4426.

Rojas, M., I. Cascorbi, K. Alexandrov, E. Kriek, G. Auburtin, L. Mayer, A. Kopp-Schneider, I. Roots, and H. Bartsch. 2000. Modulation of Benzo(a)pyrene Diolepoxide-DNA Adduct Levels in Human White Blood Cells by CYP1A1, GSTM1 and GSTT1 Polymorphism. *Carcinogenesis,* 21(1):35-41.

Romach, E.H., C.Q. Zhao, L.M. Del Razo, M.E. Cebrian, and M.P. Waalkes. 2000. Studies on the Mechanisms of Arsenic-Induced Self Tolerance Developed in Liver Epithelial Cells Through Continuous Low-level Arsenite Exposure. *Toxicol. Sci.,* 54(2):500-508.

Rossman, T.G. 1998. Molecular and Genetic Toxicology of Arsenic. In *Environmental Toxicology: Current Developments*. Edited by Rose, J. Amsterdam: Gordon and Breach Science Publishers.

Rossman, T.G., E.I. Goncharova, T. Rajah, and Z. Wang. 1997. Human Cells Lack the Inducible Tolerance to Arsenite Seen in Hamster Cells. *Mutat. Res.,* 386:307-314.

Rossman, T.G., D. Stone, M. Molina, and W. Troll. 2001. Arsenite is a Co-carcinogen With Solar Radiation for Mouse Skin: An Animal Model for Arsenic Carcinogenesis. *Toxicol. Appl. Pharmacol.,* 176:64-71.

Rotman, G., and Y. Shiloh. 1999. ATM: A Mediator of Multiple Responses to Genotoxic Stress. *Oncogene,* 18:6135-6144.

Rubin, H. 2001. Synergistic Mechanisms in Carcinogenesis by Polycyclic Aromatic Hydrocarbons and by Tobacco Smoke: A Biohistorical Perspective With Updates. *Carcinogenesis,* 22:1903-1930.

Rubin, H. 2002. Selective Clonal Expansion and Microenvironmental Permissiveness in Tobacco Carcinogenesis. *Oncogene,* 21:7392-7411.

Sampayo-Reyes, A., R.A. Zakharyan, S.M. Healy, and H.V. Aposhian. 2000. Monomethylarsonic Acid Reductase and Monomethylarsonous Acid in Hamster Tissue. *Chem. Res. Toxicol.,* 13:1181-1186.

Sancar, A. 1994. Mechanisms of DNA Excision Repair. *Science,* 266(5193):1954-1956.

Sancar, A. 1996. DNA Excision Repair. *Annu. Rev. Biochem.,* 65:43-81.

Sancar, A. 1999. Excision Repair Invades the Territory of Mismatch Repair. *Nature,* 21:247.

Sano, T., C.L. Smith, and C.R. Cantor. 1992. Immuno-PCR Very Sensitive Antigen-Detection by Means of Specific Antibody-DNA Conjugates. *Science,* 258(5079):120-122.

Santella, R.M. 1999. Immunological Methods for Detection of Carcinogen-DNA Damage in Humans. *Cancer Epidemiol. Biomarkers Prevent.,* 8:733-739.

Santella, R.M., C.D. Lin, W.L. Cleveland, and I.B. Weinstein. 1984. Monoclonal Antibodies to DNA Modified by a Benzo(a)pyrene Diol Epoxide. *Carcinogenesis,* 5(3):373-377.

Santella, R.M., F.P. Perera, T.L. Young, Y.J. Zhang, S. Chiamprasert, D. Tang, L.W. Wang, A. Beachman, J.H. Lin, and V.A. DeLeo. 1995. Polycyclic Aromatic Hydrocarbon-DNA and Protein Adducts in Coal Tar-Treated Patients and Controls and Their Relationship to Glutathione *S*-transferase Genotype. *Mutat. Res.,* 334:117-124.

Santella, R.M., A. Weston, F.P. Perera, G.T. Trivers, C.C. Harris, T.L. Young, D. Nguyen, B.M. Lee, and M.C. Poirier. 1988. Interlaboratory Comparison of Antisera and Immunoassays for Benzo(a)pyrene-Diol-Epoxide-I Modified DNA. *Carcinogenesis,* 9(7):1265-1269.

Schaumloffel, N., and T. Gebel. 1998. Heterogeneity of the DNA Damage Provoked by Antimony and Arsenic. *Mutagenesis,* 13:281-286.

Schmalzing, D., W. Nashabeh, X.W. Yao, R. Mhatre, F.E. Regnier, N.B. Afeyan, and M. Fuchs. 1995. Capillary Electrophoresis-Based Immunoassay for Cortisol in Serum. *Anal. Chem.*, 67(3):606-612.

Schultz, N.M., L. Huang, and R.T. Kennedy. 1995. Capillary Electrophoresis Based Immunoassay to Determine Insulin Content and Insulin-Secretion From Single Islets of Langerhans. *Anal. Chem.*, 67(5):924-929.

Sedelnikova, O.A., E.P. Rogakou, I.G. Panyutin, and W.M. Bonner. 2002. Quantitative Detection of (125) IdU-Induced DNA Double-Strand Breaks With Gamma-H2AX Antibody. *Radiat. Res.*, 158(4):486-492.

Seidel, C.A.M., A. Schulz, and M.H.M. Sauer. 1996. Nucleobase-Specific Quenching of Fluorescent Dyes. 1. Nucleobase One-Electron Redox Potentials and Their Correlation With Static and Dynamic Quenching Efficiencies. *J. Phys. Chem.*, 100(13):5541-5553.

Serves, S.V., Y.C. Charalambidis, D.N. Sotiropoulos, and P.V. Ioannou. 1995. Reaction of Arsenic(III) Oxide, Arsenous and Arsenic Acids With Thiols. *Phosphorus Sulfur Silicon Relat. Elem.*, 105(1-4):109-116.

Shackelford, R.E., C.L. Innes, S.O. Sieber, A.N. Heinloth, S.A. Leadon, and R.S. Paules. 2001. The Ataxia Telangiectasia Gene Product is Required for Oxidative Stress-Induced G1 and G2 Checkpoint Function in Human Fibroblasts. *J. Biol. Chem.*, 276(24):21951-21959.

Sharma, M., R. Jain, E. Ionescu, and H.K. Slocum. 1995. Capillary Electrophoretic Separation and Laser-Induced Fluorescence Detection of the Major DNA Adducts of Cisplatin and Carboplatin. *Anal. Biochem.*, 228:307-311.

Shigenaga, M.K., and B.N. Ames. 1991. Assays for 8-hydroxy-2'-deoxyguanosine: A Biomarker of *in vivo* Oxidative DNA Damage. *Free Radic. Biol. Med.*, 10(3-4):211-216.

Shimizu, M., J.F. Hochadel, B.A. Fulmer, and M.P. Waalkes. 1998. Effect of Glutathione Depletion and Metallothionein Gene Expression on Arsenic-Induced Cytotoxicity and c-myc Expression *in vitro*. *Toxicol. Sci.*, 45:204-211.

Singh, N.P., M.T. McCoy, R.R. Tice, and E.L. Schneider. 1988. A Simple Technique for Quantitation of Low Levels of DNA Damage in Individual Cells. *Exp. Cell Res.*, 175(1):184-191.

Singh, N.P., R.E. Stephens, and E.L. Schneider. 1994. Modifications of Alkaline Microgel Electrophoresis for Sensitive Detection of DNA Damage. *Int. J. Radiat. Biol.*, 66(1):23-28.

Smith, A.H., M.L. Biggs, L. Moore, R. Haque, C. Steinmaus, J. Chung, A. Hernandez, and P. Lopipero. 1999. Cancer Risks From Arsenic in Drinking Water: Implications for Drinking Water Standards. In *Arsenic Exposure and Health Effects*. Edited by Chappell, W.R., C.O. Abernathy, and R.L. Calderon. Amsterdam: Elsevier.

Smith, A.H., M. Goycolea, R. Haque, and M.L. Biggs. 1998. Marked Increase in Bladder and Lung Cancer Mortality in a Region of Northern Chile due to Arsenic in Drinking Water. *Am. J. Epidemiol.*, 147:660-669.

Smith, A.H., C. Hopenhayn-Rich, M.N. Bates, H.M. Goeden, I. Hertz-Picciotto, H.M. Duggan, R. Wood, M.J. Kosnett, and M.T. Smith. 1992. Cancer Risks From Arsenic in Drinking Water. *Environ. Health Perspect.*, 97:259-267.

Snyder, L.R., and J.J. Kirkland. 1977. *Introduction to Modern Liquid Chromatography*. 2nd ed. New York: John Wiley.

Snyder, R.D., G.F. Davis, and P.J. Lachmann. 1989. Inhibition by Metals of X-Ray and Ultraviolet-Induced DNA Repair in Human Cells. *Biol. Trace Elem. Res.,* 21:389-398.

Sokal, R.R., and F.J. Rohlf. 1969. *Biometry: The Principles and Practice of Statistics in Biological Research.* San Francisco: W.H. Freeman.

Stevens, S.Y., W. Hu, T. Gladysheva, B.P. Rosen, E.R.P. Zuiderweg, and L. Lee. 1999. Secondary Structure and Fold Homology of the ArsC Protein From the *Escherichia coli* Arsenic Resistance Plasmid R773. *Biochemistry,* 38(31):10178-10186.

Stevenson, K.J., G. Hale, and R.N. Perham. 1978. Inhibition of Pyruvate-Dehydrogenase Multi-Enzyme Complex From *Escherichia coli* With Monofunctional and Bifunctional Arsenoxides. *Biochemistry,* 17:2189-2192.

Stillman, M.J., C.F. Shaw, III, and K.T. Suzuki. 1992. *Metallothioneins: Synthesis, Structure, and Properties of Metallothioneins, Phytochelatins and Metal-Thiolate Complexes.* New York: VCH.

Stillman, M.J., D. Thomas, C. Trevithick, X. Guo, and M. Siu. 2000. Circular Dichroism, Kinetic and Mass Spectrometric Studies of Copper(I) and Mercury(II) Binding to Metallothionein. *J. Inorg. Biochem.,* 79(1-4):11-19.

Styblo, M., M. Delnomdedieu, and D.J. Thomas. 1995. Biological Mechanisms and Toxicological Consequences of the Methylation of Arsenic. In *Toxicology of Metals - Biochemical Aspects, Handbook of Experimental Pharmacology.* Edited by Goyer, R.A., and G. Cherian. Berlin: Springer-Verlag.

Styblo, M., M. Delnomdedieu, and D.J. Thomas. 1996. Mono- and Dimethylation of Arsenic in Rat Liver Cytosol *in vitro. Chem. Biol. Interact.,* 99:147-164.

Styblo, M., S.V. Serves, W.R. Cullen, and D.J. Thomas. 1997. Comparative Inhibition of Yeast Glutathione Reductase by Arsenicals and Arsenothiols. *Chem. Res. Toxicol.,* 10:27-33.

Styblo, M., and D.J. Thomas. 1995. *In vitro* Inhibition of Glutathione Reductase by Arsenotriglutathione. *Biochem. Pharmacol.,* 49:971-974.

Styblo, M., and D.J. Thomas. 1997. Binding of Arsenicals to Proteins in an *in vitro* Methylation System. *Toxicol. Appl. Pharmacol.,* 147:1-8.

Styblo, M., L. Vega, D.R. Germolec, M.I. Luster, L.M. Del Razo, C. Wang, W.R. Cullen, and D.J. Thomas. 1999. Metabolism and Toxicity of Arsenicals in Cultured Cells. In *Arsenic Exposure and Health Effects.* Edited by Chappell, W.R., C.O. Abernathy, and R.L. Calderon. Amsterdam: Elsevier.

Styblo, M., H. Yamauchi, and D.J. Thomas. 1995. Comparative *in vitro* Methylation of Trivalent and Pentavalent Arsenicals. *Toxicol. Appl. Pharmacol.,* 135:172-178.

Su, Y., and M. Swift. 2000. Mortality Rates Among Carriers of Ataxia-Telangiectasia Mutant Alleles. *Ann. Intern. Med.,* 133:770-778.

Szeliga, J., and A. Dipple. 1998. DNA Adduct Formation by Polycyclic Aromatic Hydrocarbon Dihydrodiol Epoxides. *Chem. Res. Toxicol.,* 11(1):1-11.

Tam, G.K.H., S.M. Charbonneau, F. Bryce, C. Pomroy, and E. Sandi. 1979. Metabolism of Inorganic Arsenic ($^{74}$As) in Humans Following Oral Ingestion. *Toxicol. Appl. Pharmacol.,* 50:319-322.

Tan, W.G., T.J. Carnelley, P. Murphy, H. Wang, J. Lee, S. Barker, M. Weinfeld, and X.C. Le. 2001. Detection of DNA Adducts of Benzo(a)pyrene Using Immuno-Electrophoresis With Laser-Induced Fluorescence: Analysis of A549 Cells. *J. Chromatogr. A,* 924:377-386.

Tang, M., A. Pao, and X. Zhang. 1994. Repair of Benzo(a)pyrene Diol Epoxide- and UV-Induced DNA Damage in Dihydrofolate Reductase and Adenine Phosphoribosyltransferase Genes of CHO Cells. *J. Biol. Chem.,* 269(17):12749-12754.

Tao, L., and R.T. Kennedy. 1996. On-line Competitive Immunoassay for Insulin Based on Capillary Electrophoresis With Laser-Induced Fluorescence Detection. *Anal. Chem.,* 68:3899-3906.

Thames, H.D., Jr., and S.L. Rasmussen. 1978. A Test for Dose-Modifying Factors. *Radiat. Res.,* 76:308-324.

Thomale, J., J. Engelbergs, F. Seiler, and M.F. Rajewsky. 1996. Monoclonal Antibody-Based Quantification and Repair Analysis of Specific Alkylation Products in DNA. In *Technologies for Detection of DNA Damage and Mutations*. Edited by Pfeifer, G.P. New York: Plenum Press.

Thompson, D.J. 1993. A Chemical Hypothesis for Arsenic Methylation in Mammals. *Chem. Biol. Interact.,* 88(2-3):89-114.

Thorgiersson, U.P., D.W. Dalgard, J. Reeves, and R.H. Adamson. 1994. Tumour Incidence in a Chemical Carcinogenesis Study of Nonhuman Primates. *Regul. Toxicol. Pharmacol.,* 19:130-151.

Thrall, B.D., D.B. Mann, M.J. Smerdon, and D.L. Springer. 1992. DNA Polymerase, RNA Polymerase and Exonuclease Activities on a DNA Sequence Modified by Benzo(a)pyrene Diolepoxide. *Carcinogenesis,* 13:1529-1534.

Tondel, M., M. Rahman, A. Magnuson, I.A. Chowdhury, M.H. Faruqee, and S.A. Ahmad. 1999. The Relationship of Arsenic Levels in Drinking Water and the Prevalence Rate of Skin Lesions in Bangladesh. *Environ. Health Perspect.,* 107(9):727-729.

Toyama, M., M. Yamashita, N. Hirayama, and Y. Murooka. 2002. Interactions of Arsenic With Human Metallothionein-2. *J. Biochem.,* 132:217-221.

Tran, H.P., A.S. Prakash, R. Barnard, B. Chiswell, and J.C. Ng. 2002. Arsenic Inhibits the Repair of DNA Damage Induced by Benzo(a)pyrene. *Toxicol. Lett.,* 133:59-67.

Tseng, C.H., T.Y. Tai, C.K. Chong, C.P. Tseng, M.S. Lai, B.J. Lin, H.Y. Chiou, Y.M. Hsueh, K.H. Hsu, and C.J. Chen. 2000. Long-term Arsenic Exposure and Incidence of Non-Insulin-Dependent Diabetes Mellitus: A Cohort Study in Arseniasis in Hyperendemic Villages in Taiwan. *Environ. Health Perspect.,* 108:847-851.

Tseng, W.P. 1977. Effects and Dose-Response Relationships of Skin Cancer and Blackfoot Disease With Arsenic. *Environ. Health Perspect.,* 19:109-119.

Tsuda, T., A. Babazono, E. Yamamoto, N. Kurumatani, Y. Mino, T. Ogawa, Y. Kishi, and H. Aoyama. 1995. Ingested Arsenic and Internal Cancer: A Historical Cohort Study Followed for 33 Years. *Am. J. Epidemiol.,* 141:198-209.

Vahter, M. 1983. Metabolism of Arsenic. In *Biological and Environmental Effects of Arsenic*. Edited by Fowler, B.A. Amsterdam: Elsevier.

Vahter, M. 1994a. Species Differences in the Metabolism of Arsenic Compounds. *Appl. Organomet. Chem.,* 8:175-182.

Vahter, M. 1994b. What are the Chemical Forms of Arsenic in the Urine, and What can They Tell Us About Exposure? *Clin. Chem.,* 40:679-680.

Vahter, M. 1999. Variation in Human Metabolism of Arsenic. In *Arsenic Exposure and Health Effects*. Edited by Chappell, W.R., C.O. Abernathy, and R.L. Calderon. Amsterdam: Elsevier.

Vahter, M., G. Concha, B. Nermell, R. Nilsson, F. Dulout, and A.T. Natarajan. 1995a. A Unique Metabolism of Inorganic Arsenic in Native Andean Women. *Eur. J. Pharmacol. Environ. Toxicol.,* 293:455-462.

Vahter, M., R. Couch, B. Nermell, and R. Nilsson. 1995b. Lack of Methylation of Inorganic Arsenic in the Chimpanzee. *Toxicol. Appl. Pharmacol.,* 133:262-268.

Vahter, M., and E. Marafante. 1983. Intracellular Interaction and Metabolic Fate of Arsenite and Arsenate in Mice and Rabbits. *Chem. Biol. Interact.,* 47:29-44.

Vahter, M., and E. Marafante. 1985. Reduction and Binding of Arsenate in Marmoset Monkeys. *Arch. Toxicol.,* 57:119-124.

Vakharia, D.D., N. Liu, R. Pause, M. Fasco, E. Bessette, Q.Y. Zhang, and L.S. Kaminsky. 2001. Polycyclic Aromatic Hydrocarbon/Metal Mixtures: Effect on PAH Induction of CYP1A1 in Human HEPG2 Cells. *Drug Metab. Dispos.,* 29:999-1006.

Valles Mota, J.P., A.R. Linde Arias, M.R. Fernandez de la Campa, J.I. Garcia Alonso, and A. Sanz-Medel. 2000. Development of a Stable Isotope Approach for the Inductively Coupled Plasma-Mass Spectrometry Determination of Oxidized Metallothionein in Biological Materials. *Anal. Biochem.,* 282(2):194-199.

Vamosi, G., and R.M. Clegg. 1998. The Helix-Coil Transition of DNA Duplexes and Hairpins Observed by Multiple Fluorescence Prameters. *Biochemistry,* 37:14300-14316.

Venkatachalam, S., M.F. Denissenko, and A.A. Wani. 1995. DNA Repair in Human Cells: Quantitative Assessment of Bulky *anti*-BPDE-DNA Adducts by Non-competitive Immunoassays. *Carcinogenesis,* 16(9):2029-2036.

Vogt, B.L., and T.G. Rossman. 2001. Effects of Arsenite on p53, p21 and Cyclin D Expression in Normal Human Fibroblasts – A Possible Mechanism for Arsenite's Comutagenicity. *Mutat. Res.,* 478(1-2):159-168.

Waalkes, M.P., M.J. Harvey, and C.D. Klaassen. 1984. Relative *in vitro* Affinity of Hepatic Metallothionein for Metals. *Toxicol. Lett.,* 20:33-39.

Wan, Q.H., and X.C. Le. 1999. Capillary Electrophoresis Coupled With Laser-Induced Fluorescence Polarization as a Hybrid Approach to Ultrasensitive Immunoassays. *J. Chromatogr. A,* 853:555-562.

Wan, Q.H., and X.C. Le. 2000. Studies of Protein-DNA Interactions by Capillary Electrophoresis Laser-Induced Fluorescence Polarization. *Anal. Chem.,* 72:5583-5589.

Wanabuchi, H., S. Yamamoto, H. Chen, K. Yoshida, G. Endo, T. Hori, and S. Fukushima. 1996. Promoting Effects of Dimethylarsinic Acid on *N*-butyl-*N*-(4-hydroxybutyl) Nitrosamine-Induced Urinary Bladder Carcinogenesis in Rats. *Carcinogenesis,* 17:2435-2439.

Wang, H., M. Lu, N. Mei, J. Lee, M. Weinfeld, and X.C. Le. 2003. Immunoassays using capillary electrophoresis with laser induced fluorescence detection for DNA adducts. *Anal. Chim. Acta,* 500: in press.

Wang, H., and X.C. Le. N.d. The Focusing of Protein-DNA Complexes in Capillary Electrophoresis Based on DNA Binding Assay. In preparation.

Wang, H., M. Lu, M. Weinfeld, and X.C. Le. 2003. Enhancement of Immunocomplex Detection and Application to Assays for DNA Adducts of Benzo(a)pyrene. *Anal. Chem.,* 75(2):247-254.

Wang, H., J. Xing, W. Tan, M. Lam, T. Carnelley, M. Weinfeld, and X.C. Le. 2002. Binding Stoichiometry of DNA Adducts With Antibodies Studied by Capillary Electrophoresis and Laser-Induced Fluorescence. *Anal. Chem.,* 74(15):3714-3719.

Wang, J., G. Rivas, M. Ozsos, D.H. Grant, X.H. Cai, and C. Parrado. 1997. Microfabricated Electrochemical Sensor for the Detection of Radiation-Induced DNA Damage. *Anal. Chem.,* 69(7):1457-1460.

Wang, T.S., T.Y. Hsu, C.H. Chung, A.S. Wang, D.T. Bau, and K.Y. Jan. 2001. Arsenite Induces Oxidative DNA Adducts and DNA-Protein Cross-Links in Mammalian Cells. *Free Radic. Biol. Med.,* 31:321-330.

Wani, M.A., Q. Zhu, M. El-Mahdy, S. Venkatachalam, and A.A. Wani. 2000. Enhanced Sensitivity to *anti*-benzo(a)pyrene-diol-epoxide DNA Damage Correlates With Decreased Global Genomic Repair Attributable to Abrogated p53 Function in Human Cells. *Cancer Res.,* 60(8):2273-2280.

Ward, J.F. 1994. The Complexity of DNA Damage: Relevance to Biological Consequences. *Int. J. Radiat. Biol.,* 66:427-432.

Weinfeld, M., and G.W. Buchko. 1993. Postlabeling Methods for the Detection of Apurinic Sites and Radiation-Induced DNA Damage. In *Postlabeling Methods for the Detection of DNA Damage*. Edited by Phillips, D.H., M. Castegnaro, and H. Bartsch. Lyon: IARC Scientific Publications.

Weinfeld, M., J. Lee, G. Ruiqi, F. Karimi-Busheri, D. Chen, and J. Allalunis-Turner. 1997. Use of a Postlabeling Assay to Examine the Removal of Radiation-Induced DNA Lesions by Purified Enzymes and Human Cell Extracts. *Mutat. Res.,* 378:127-137.

West, G.J., I.W.L. West, and J.F. Ward. 1982. Radioimmunoassay of a Thymine Glycol. *Radiat. Res.,* 90(3):595-608.

Weston, A. 1993. Physical Methods for the Detection of Carcinogen-DNA Adducts in Humans. *Mutat. Res.,* 288:19-29.

Weston, A., and E.D. Bowman. 1991. Fluorescence Detection of Benzo(a)pyrene-DNA Adducts in Human Lung. *Carcinogenesis,* 12(8):1445-1449.

Whitehouse, C.J., R.M. Taylor, A. Thistlethwaite, H. Zhang, F. Karimi-Busheri, D.D. Lasko, M. Weinfeld, and K.W. Caldecott. 2001. XRCC1 Stimulates Human Polynucleotide Kinase Activity at Damaged DNA Termini and Accelerates DNA Single-Strand Break Repair. *Cell,* 104:107-117.

WHO (World Health Organization). 2001. *Environmental Health Criteria 224: Arsenic and Arsenic Compounds*. Geneva: World Health Organization.

Wiencke, J.K., J.W. Yager, A. Varkonyi, M. Hultner, and L.H. Lutze. 1997. Study of Arsenic Mutagenesis Using the Plasmid Shuttle Vector pZ189 Propagated in DNA Repair Proficient Human Cells. *Mutat. Res.,* 386:335-344.

Wolf, S.M., and P. Vouros. 1994. Application of Capillary Liquid Chromatography Coupled With Tandem Mass-Spectrometric Methods to the Rapid Screening of Adducts Formed by the Reaction of N-acetoxy-N-acetyl-2-aminofluorene With Calf Thymus DNA. *Chem. Res. Toxicol.,* 7(1):82-88.

Wood, R.D. 1996. DNA Repair in Eukaryotes. *Annu. Rev. Biochem.,* 65:135-167.

Wood, R.D., P. Robins, and T. Lindahl. 1988. Complementation of the Xeroderma Pigmentosum DNA Repair Defect in Cell-Free Extracts. *Cell,* 53:97-106.

Xing, J.Z., T. Carnelley, J. Lee, W.P. Watson, E. Booth, M. Weinfeld, and X.C. Le. 2001. Assay for DNA Damage Using Immunochemical Recognition and Capillary Electrophoresis. *Methods Mol. Biol.,* 162:419-428.

Xing, J.Z., J. Lee, S.A. Leadon, M. Weinfeld, and X.C. Le. 2000. Measuring DNA Damage Using Capillary Electrophoresis With Laser-Induced Fluorescence Detection. *Methods,* 22(2):157-163.

Yager, J.W., J.B. Hicks, and E. Fabianova. 1997. Airborne Arsenic and Urinary Excretion of Arsenic Metabolites During Boiler Cleaning Operations in a Slovak Coal-Fired Power Plant. *Environ. Health Perspect.,* 105:836-842.

Yamamoto, A., A. Hisanaga, and N. Ishinishi. 1987. Tumorigenicity of Inorganic Arsenic Compounds Following Intratracheal Instillations to the Lungs of Hamsters. *Int. J. Cancer,* 40(2):220-223.

Yamamoto, S., Y. Konishi, T. Matsuda, T. Murai, M.A. Shibata, I. Matsuiyuasa, S. Otani, K. Kuroda, G. Endo, and S. Fukushima. 1995. Cancer Induction by an Organic Arsenic Compound, Dimethylarsinic Acid (Cacodylic Acid), in F344/Ducrj Rats after Pretreatment With 5 Carcinogens. *Cancer Res.,* 55(6):1271-1276.

Yamauchi, H., and B.A. Fowler. 1994. Toxicity and Metabolism of Inorganic and Methylated Arsenicals. In *Arsenic in the Environment, Part II: Human Health and Ecosystem Effects*. Edited by Nriagu, J.O. New York: Wiley.

Yamauchi, H., K. Takahashi, M. Mashiko, J. Saitoh, and Y. Yamamura. 1992. Intake of Different Chemical Species of Dietary Arsenic by the Japanese, and Their Blood and Urinary Arsenic Levels. *Appl. Organomet. Chem.,* 6:383-388.

Yamauchi, H., K. Takahashi, M. Mashiko, and Y. Yamamura. 1989. Biological Monitoring of Arsenic Exposure of Gallium Arsenide- and Inorganic Arsenic-Exposed Workers by Determination of Inorganic Arsenide and its Metabolites in Urine and Hair. *Am. Ind. Hyg. Assoc. J.,* 50:606-612.

Yates, J.R., 3rd, J.K. Eng, and A.L. McCormack. 1995. Mining Genomes: Correlating Tandem Mass Spectra of Modified and Unmodified Peptides to Sequences in Nucleotide Databases. *Anal. Chem.,* 67(18):3202-3210.

Ye, L., X.C. Le, J.Z. Xing, M. Ma, and R. Yatscoff. 1998. Competitive Immunoassay for Cyclosporine Using Capillary Electrophoresis With Laser Induced Fluorescence Polarization Detection. *J. Chromatogr. B,* 714:59-67.

Yih, L.H., and T.C. Lee. 2000. Arsenite Induces p53 Accumulation Through an ATM-Dependent Pathway in Human Fibroblasts. *Cancer Res.,* 60:6346-6352.

Yu, X., M. Wojciechowski, and C. Fenselau. 1993. Assessment of Metals in Reconstituted Metallothioneins by Electrospray Mass Spectrometry. *Anal. Chem.,* 65(10):1355-1359.

Zaia, J., D. Fabris, D. Wei, R.L. Karpel, and C. Fenselau. 1998. Monitoring Metal Ion Flux in Reactions of Metallothionein and Drug-Modified Metallothionein by Electrospray Mass Spectrometry. *Protein Sci.,* 7(11):2398-2404.

Zakharyan, R.A., F. Ayala-Fierro, W.R. Cullen, D.M. Carter, and H.V. Aposhian. 1999. Enzymatic Methylation of Arsenic Compounds. VII. Monomethylarsonous Acid ($MMA^{III}$) is the Substrate for MMA Methyltransferase of Rabbit Liver and Human Hepatocytes. *Toxicol. Appl. Pharmacol.,* 158:9-15.

Zakharyan, R.A., Y. Wu, G.M. Bogdan, and H.V. Aposhian. 1995. Enzymatic Methylation of Arsenic Compounds: Assay, Partial Purification, and Properties of Arsenite Methyltransferase and Monomethylarsonic Acid Methyltransferase of Rabbit Liver. *Chem. Res. Toxicol.,* 8:1029-1038.

Zeisig, M., and L. Moller. 1995. 32P-HPLC Suitable for Characterization of DNA Adducts Formed *in vitro* by Polycyclic Aromatic Hydrocarbons and Derivatives. *Carcinogenesis,* 16:1-9.

Zhang, Z.H., and M.C. Poirier. 1997. Cisplatin-DNA Adduct Determination in the Hepatic Albumin Gene as Compared to Whole Genomic DNA. *Chem. Res. Toxicol.,* 10(9):971-977.

Zhao, C.Q., M.R. Young, B.A. Diwan, T.P. Coogan, and M.P. Waalkes. 1997. Association of Arsenic-Induced Malignant Transformation With DNA Hypomethylation and Aberrant Gene Expression. *Proc. Natl. Acad. Sci. USA,* 94:10907-10912.

Zhao, S., Y.C. Weng, S.S. Yuan, Y.T. Lin, H.C. Hsu, S.C. Lin, E. Gerbino, M.H. Song, M.Z. Zdzienicka, R.A. Gatti, J.W. Shay, Y. Ziv, Y. Shiloh, and E.Y. Lee. 2000. Functional Link Between Ataxia-Telangiectasia and Nijmegen Breakage Syndrome Gene Products. *Nature,* 405:473-477.

Zhong, C.X., and M.J. Mass. 2001. Both Hypomethylation and Hypermethylation of DNA Associated With Arsenite Exposure in Cultures of Human Cells Identified by Methylation-Sensitive Arbitrarily-Primed PCR. *Toxicol. Lett.,* 122(3):223-234.

Ziv, Y., A. Bar-Shira, I. Pecker, P. Russell, T.J. Jorgensen, I. Tsarfati, and Y. Shiloh. 1997. Recombinant ATM Protein Complements the Cellular A-T Phenotype. *Oncogene,* 15:159-167.

## ABBREVIATIONS AND GLOSSARY

AP lyase: In BER, a DNA cleavage activity associated with some DNA glycosylases in which phosphodiester bond breakage occurs by β-elimination 3' to an apurinic/apyrimidinic (AP) site.

AP-1: Transcription factor activator protein 1 is an important regulator of nuclear gene expression.

$As^{III}$: Inorganic arsenite, $As(OH)_3$

$As^{V}$: Inorganic arsenate, $AsO(OH)_3$

BaP: Benzo[a]pyrene is a polycyclic aromatic hydrocarbon which may be metabolized to highly carcinogenic electrophilic epoxide (e.g. BPDE)

BER: Base Excision Repair is a mode of DNA repair in which nonbulky DNA adducts such as thymine glycols are removed. A DNA glycosylase cleaves the altered base from the deoxyribose, then the apurinic/apyrimidinic (AP) sugar is released by the combined actions of AP lyase and an AP endonuclease. (see Figure 1)

BPDE: Benzo[a]pyrene diol-epoxide is a highly carcinogenic metabolite of BaP which can bind to DNA, particularly guanine residues, to generate bulky adducts.

BrdU: Bromodeoxyuridine is a thymidine analog which in its enol form exhibits base-pairing properties similar to cytosine, thus potentiating transversions in the DNA sequence.

CE: Capillary electrophoresis, an electrophoretic separation technique carried out using small capillaries as a separation column.

$DMA^{III}$: Dimethylarsinous acid, $(CH_3)_2As(OH)$

$DMA^{V}$: Dimethylarsinic acid, $(CH_3)_2AsO(OH)$

DMPS: 2,3dimercapto-1-propane sulfonate, a chelating agent

Endonuclease: An enzyme capable of nucleotidyl hydrolytic cleavage of DNA by hydrolyzing the C(3')-O-P bond of the DNA backbone, leaving 3'-hydroxyl-nucleotide and 5'-deoxyribose-phosphate termini on the cleaved strand.

ERCC1: Excision repair cross complementing protein 1 interacts with XPF and other proteins to form the complex which performs incision of the DNA strand in NER

HAP1 (Ape): Human AP endonuclease enzyme acts in BER to cleave DNA strands at abasic (AP) sites.

hNth: Human thymine-glycol-DNA glycosylase acts at an early stage in BER to recognize and excise damaged pyrimidine residues. This combined DNA glycosylase and apyrimidinic lyase is the structural and functional human homologue of *E. coli* endonuclease III ($nth^{+}$), and has thus also been named hNTH1.

HG: Hydride generation, a derivatization technique that is used to convert arsenic species in solution to volatile arsines.

HGAFS: Hydride generation atomic fluorescence spectrometry, a detection technique

HPLC: High performance liquid chromatography, a separation technique

ICPMS: Inductively coupled plasma mass spectrometry, a technique commonly used for trace element analysis.

LIF: Laser induced fluorescence, a technique that detects fluorescence of compounds being excited by a laser

$MMA^{III}$: Monomethylarsonous acid, $CH_3As(OH)_2$

$MMA^{V}$: Monomethylarsonic acid, $CH_3AsO(OH)_2$

MT: Metallothionein, a small protein rich in cysteine residues

NER: Nucleotide Excision Repair is a mode of DNA repair in which bulky adducts such as cyclobutane pyrimidine dimers and benzo(a)pyrene diol epoxide adducts are removed. An enzyme system hydrolyzes two phosphodiester bonds some distance on either side of the adduct, releasing an oligonucleotide bearing the bulky adduct, and the gap on the DNA is filled in by a polymerase and ligated to repair the strand.

Northern blot: A process in which a mixture of RNA molecules are separated by gel electrophoresis, bound to nitrocellulose membrane, and specific RNA(s) identified by their ability to hybridize to radiolabelled DNA or RNA probes.

PAH: Polycyclic aromatic hydrocarbon

PBS: Phosphate buffered saline

PARP: Poly(ADP-ribose) polymerase binds to single-strand breaks and acts as a critical regulatory component in BER. PARP may recruit BER mechanisms to the repair of strand interruptions, and also enhances transcription by activation of the preinitiation complex during its formation.

PCNA: Proliferating cell nuclear antigen is a ring-shaped homotrimeric protein that encircles DNA acting as a polymerase accessory factor, and is necessary for DNA replication and for resynthesis of damaged DNA during NER.

polymerase: An enzyme which catalyzes the incorporation of free nucleotides in the chain-elongation of a DNA strand from a oligonucleotide primer along a single-stranded DNA template.

RPA: Human replication protein, a heterotrimeric single-stranded DNA binding protein, is one of the most abundant proteins in the cell, with functions in DNA replication, recombination and repair. In NER it forms an association with XPA and XPG, perhaps targeting these proteins to the site of bulky DNA adducts.

SAM: S-adenosylmethionine, a methyl donor

TFIIH: Transcription factor IIH is required for accurate transcription of protein-coding genes by RNA polymerase II, and has an essential role in NER. The protein products of the *XPB* and *XPD* genes are two of the nine subunits which comprise this complex, and are involved with clearing the transcription machinery from the promoter region following its assembly.

Tg: Thymine glycols are nonbulky DNA lesions generated by the oxidative damage to DNA in the thymine base.

TMA: Trimethylarsine oxide, $(CH_3)_3As$

TMAO: Trimethylarsine oxide, $(CH_3)_3AsO$

Western blot: A process in which a mixture of proteins is separated by gel electrophoresis, bound to nitrocellulose membrane, and specific protein(s) identified by their ability to bind specific (often radiolabeled or fluorescent labeled) antibodies.

XP: Xeroderma pigmentosum is a human disorder involving deficiencies in DNA repair, in particular the nucleotide excision repair (NER) pathway. Patients have been grouped into seven different complementation groups (XP-A to XP-G), which are correlated to seven different genes (*XPA* to *XPG*). The protein products (XPA to XPG) of these genes perform essential tasks in NER, and several also serve other roles (e.g. XPG in BER; XPB and XPD in transcription machinery assembly).

IWA Publishing's authorised EU representative for General Product Safety Regulations is Diane D'Arras, 15 rue Duret, 75116 Paris, France, e-mail: safety@iwap.co.uk.

Printed and bound by CPI Group (UK) Ltd, Croydon, CR0 4YY
12/05/2026
02108892-0014